D0397303

EERL
Surplus/Dup

AGRICULTURAL EXTENSION WORLDWIDE: ISSUES, PRACTICES AND EMERGING PRIORITIES

INTERNATIONAL PERSPECTIVES ON ADULT AND CONTINUING EDUCATION

Edited by Peter Jarvis, University of Surrey
Consultant Editors: Chris Duke and Ettore Gelpi

AGRICULTURAL EXTENSION WORLDWIDE

ISSUES, PRACTICES AND EMERGING PRIORITIES

Edited by
WILLIAM M. RIVERA and SUSAN G. SCHRAM

CROOM HELM
London • New York • Sydney

© 1987 W.M. Rivera
Croom Helm Ltd, Provident House, Burrell Row,
Beckenham, Kent, BR3 1AT
Croom Helm Australia, 44-50 Waterloo Road,
North Ryde, 2113, New South Wales

British Library Cataloguing in Publication Data

Agricultural extension worldwide: issues,
    practices and emerging priorities. —
    (Croom Helm series in international adult
    education)
    1. Agricultural extension work
    I. Rivera, W.M. II. Schram, Susan G.
    630'.7'15   S544
    ISBN 0-7099-4238-9

Published in the USA by
Croom Helm
in association with Methuen, Inc.
29 West 35th Street
New York, NY 10001

Library of Congress Cataloging-in-Publication Data

ISBN 0-7099-4238-9

Printed and bound in Great Britain
by Billing & Sons Limited, Worcester.

CONTENTS

Acknowledgements

Introduction   1

I ISSUES

EDITOR'S NOTE

The Croom Helm Series in International Adult Education brings to an English-speaking readership a wide overview of developments in the Education of Adults throughout the world. Books published and others that are planned in this series of at least four different types:

a) those concerning adult and continuing education in a single country,
b) those having a comparative perspective of two or more countries,
c) studies having an international perspective,
d) symposia from different countries having a single theme.

The present book has a single theme, which will be familiar to adult educators who work in agricultural extension - one of the largest single areas of adult education worldwide - but which is less familiar to others in the field. In this book, the authors explore the policies, issues and priorities for agricultural extension, especially as it seeks to work with farmers in Third World societies to assist in development. It is important that educational policy should be examined by educators and this study will provide material for further analysis.

The papers that comprise this book emerged as a result of seminars organized under the auspices of the Center for International Extension Development at the University of Maryland, which is directed by William Rivera. They provide information, material for subsequent analysis and a special perspective on education of adults. For all of these reasons this is a welcome addition to the literature in International Adult Education.

Peter Jarvis
(Series Editor)

ACKNOWLEDGEMENTS

This volume issues from a colloquium series on the theme of 'Agricultural Extension Worldwide', organized at the University of Maryland, College Park, by the Center for International Extension Development (CIED). The CIED series was supported by the Department of Agricultural and Extension Education (AEED), the Maryland Cooperative Extension Service (MCES), the Office of International Programs (OIP) of the College of Agriculture, and the Office of International Affairs representing the College Park Campus. The former Chairman of the AEED Department, Dr Clifford L. Nelson, provided continuing support to the Center for International Extension Development, including its Agricultural Extension Worldwide Colloquium Series. Dr John R. Moore, Assistant Dean and Director of OIP, Dr John T. Rowntree, OIP Associate Coordinator, and Dr Wayne Nilsestuen, then assisting OIP (on loan during 1985 from the U.S. Agency for International Development by way of its Joint Career Corps program), helped considerably through various means of support for the Colloquium Series. Support for one of the colloquia by Dr Tal Shehata, Director of the Office of International Affairs, is also acknowledged with appreciation.

The U.S. Agency for International Development (US/AID) through its Title XII Strengthening Grant Program provided assistance to the colloquium series and toward publishing this volume. The Agency's assistance is recognized with gratitude.

The assistance of the Maryland Cooperative Extension Service's Office of Information and Publications was critical in the development of public announcements for the Series - thanks in particular to Mrs Ann Pease, Chief of Printing and Publications. Many other colleagues and graduate students, too numerous to mention, were also most helpful in hosting presenters. Special thanks to the AEED Department secretary Jeanne Smith for typing the original manuscript. It is, of course, the presenters who deserve the strongest expression of gratitude since their commitment and cooperation made this volume possible.

William M. Rivera and Susan G. Schram
University of Maryland, College Park

INTRODUCTION

This volume grew initially from the conviction of the editors that agricultural extension internationally could profit from (1) new perspectives and paradigms; (2) on-going experimentation in developing appropriate extension systems; and (3) re-invigorated vision, leadership and innovation. The volume is the result of a series of colloquia held during 1985-6 on 'Agricultural Extension Worldwide: Systems, Linkages and Supports'. The series was organized at the University of Maryland, College Park, by the Center for International Extension Development in the College of Agriculture's Department of Agricultural and Extension Education. Leading professionals in various specialities of agricultural development were asked to share their perspective on contemporary issues relating to this topic.

The editors are indebted to the contributors to this volume. The papers presented in the colloquium series greatly expanded our own understanding of the variety of forces at work, as well as the several systems and multiple processes, procedures and mechanisms involved in effectively operating agricultural extension in less-developed countries. It is hoped that this volume will be equally informative for (1) policy-makers seeking to determine the most effective extension systems to accomplish their goals; (2) international organization professionals engaged in agricultural assistance programs; (3) extension personnel internationally; and (4) faculty and students in the field.

Questions Addressed
This series began with a general understanding that: (a) agricultural extension systems are varied, with multiple approaches and models; (b) the agricultural extension system is interdependent, requiring improved linkages with other facets of the agricultural development process - with other agencies and organizations; and (c) the success of agricultural extension is determined by certain supports - in

particular economic supports as expressed through policy, planning and resource allocations -- but also including socio-cultural receptivity, sociopolitical climate, and government commitment to advance agriculture through agricultural extension programs. Thus, the series set out to explore several sets of factors external and internal to public agency agricultural extension systems that affect the successful operation and development of agricultural extension programs. It sought to deepen understanding of relationships along the continuum from the agricultural policy development arena to various individual agricultural development agencies onto the specific agency and/or system for agricultural extension and finally to the farmer and market intermediaries.

The original proposition of the series was to center discussion around the 'factors for success' operating in one or another of the following three general areas: policy, practices and program linkages. What, in each of these arenas, were the factors necessary for effective, successful extension?

Contributors were asked to illustrate their remarks with references to field experience. They were to concentrate their observations on the development of agricultural extension in those countries which are less economically developed and requiring basic needs -- what some commentators tend to categorize as the less-developed countries (LDCs) and others term 'developing countries' or the 'Third World'.

The decision to compile the colloquia manuscripts first suggested a proceedings, but with the interest of Croom Helm in publishing an edited text, the original three areas of concern which initially directed the colloquia came to be re-cast into a more interesting arrangement, organizing the sixteen chapters under the following rubrics: I. Issues; II. Practices; and III. Emerging Priorities. Discussion of different agricultural extension systems, management of their linkages with other parts of the agricultural development process, and the primary supports needed for success are nevertheless highlighted through the volume.

Several new ideas have evolved as the colloquium series and this subsequent volume have unfolded. Indeed, it now appears to us that the task of developing and changing agricultural extension services, whether in modernized or LDC countries, may require a distancing from any one agricultural extension model and a move toward the design of new paradigms (possibly integrating aspects of several models). It may also require models featuring the involvement of various agencies, as is the case already in many countries, rather than a concentration regarding extension services limited to one agency. Long-term concerns now appear to us to be with the effectiveness of knowledge transfer throughout the entire agricultural development process. The correct choice by governments as to their goals and the best system(s) for reaching these goals will of course differ from country to country and will change over time. These thoughts will be further explored in the epilogue. This volume brings together a range of specialists from several disciplines with differing perspectives. It is hoped that through diversity of perception and viewpoint may come new understandings and enlightened future actions for the improvement of extension services worldwide. Following is an editorial overview of key points presented in the various chapters. The chapters are organized under three section headings, i.e. Issues, Practices and Emerging Priorities.

ORGANIZATION OF THE VOLUME

The first six chapters of this volume address contemporary ISSUES regarding agricultural extension worldwide. William L. Rodgers, private-sector specialist at the US Agency for International Development, Latin American Bureau, discusses 'The Private Sector: Its Extension Systems and Public/Private Coordination'. Rodgers defines the range of agriculture's private sector, reminding us that it includes many different entities: individual farmers, farm enterprises, agricultural input industries, agro-service enterprises, processing industries, marketing firms, and multinational firms (and/or subsidiaries), as well as agricultural production and marketing cooperatives, farmer associations and private voluntary organizations. He emphasizes that private firms cannot substitute for public agencies and clarifies when public, private, or mixed (public/private) systems work most effectively. Rodgers notes the several motivations for private firms to become involved in agricultural extension activities, and gives examples of specific cases. Finally, he considers the value and mechanisms of public/private coordination and provides examples of US AID cooperative efforts in this regard.

Dennis A. Rondinelli, presently at the Research Triangle Institute in North Carolina, was professor at Syracuse University when he presented his comparative analysis of 'Administrative Decentralization of Agricultural and Rural Development Programs in Asia'. Rondinelli reviews nine country situations where the structure and practice of agricultural development has traditionally been highly centralized and examines policies and programs that seek to decentralize development planning and administration. Drawing on his earlier work, Decentralization and Development (Cheema and Rondinelli, 1983), Rondinelli discusses the transfer of planning, decision making, and management functions from central government to field organizations, sub-

ordinate units of government, semi-autonomous public corporations, regional development organizations, specialized
authorities, and non-governmental organizations.
   G. Edward Schuh of The World Bank contributes insights
on 'The Policy Environment Necessary to Make Extension
Effective'. His essay covers three broad topics: the policy
environment for effective extension, the changed international
environment for agriculture, and the need for economic policy
education. Schuh urges policymakers to be cognizant of the
effect of price, credit, and environmental policies which
discriminate against agriculture. He notes that developments
in the international economy have changed the context of
policy making for agriculture and underlines the linkages
between monetary flows and commodity markets. Schuh
recommends that agricultural extension systems consider
policy education to be an important component of their
mission. He stresses the complementarity among the following
areas: science and technology policies, the need for policy
education for farmers (which underlines the international as
well as national context), and the need for economic policies
that offer incentives to producers.
   Donald C. Pickering of The World Bank provides 'An
Overview of Agricultural Extension and its Linkages with
Agricultural Research', explaining in particular the African
situation and the experience of The World Bank. Following a
discussion of the design of agricultural extension systems and
various approaches to rural extension, he concludes that
successful extension must be based on a sound agricultural
development policy framework supported by appropriate
budgetary provisions. Pickering underlines five priority
considerations: cost, available technology, effective organization and planning, provision for women agriculturalists,
and the importance of continued learning (from the past as
well as from on-going activity). Stressing that there is no
'blueprint' agricultural extension system, he reviews six
lessons learned from four World Bank Workshops, held in
1984/5: (1) the need for support of farmers, local officials,
and central officials; (2) the availability of inputs; (3) the
importance of on-farm research; (4) the need for improved
collaboration among agencies; (5) the need for responsiveness
of institutions; and (6) improved management.
   Nigel Roberts of The World Bank clarifies how 'Successful Agricultural Extension' is dependent upon: (1) an agricultural research network with links to extension; (2) both
credit and input supply systems; (3) policies that provide
farmer incentive structures; and (4) effective use of government and staff. Roberts notes that agricultural extension
programs are designed in constrained environments, making
the context ultimately more important than any particular type

of system. Within that context, however, the above-mentioned four key areas must be positively operating within the agricultural development process.

J. Kenneth McDermott concludes the first section of the book with a discussion of the role of linkages between agricultural extension and research. McDermott advocates consideration of an inclusive concept that he calls the Technology Innovation Process (TIP). He maintains that there should be no clear distinction between the functions of agricultural research and extension; agricultural extension should provide 'technical liaison and support' - maintaining liaison with research and input suppliers, and providing technical support to field staff. McDermott enumerates eight components in the TIP process, and suggests that the greatest opportunity presented by the Farming Systems Research and Extension (FSR/E) approach may be its capacity to bring countries' entire technology innovation process in contact with what he calls the international technology network (ITN).

The second section deals with PRACTICES, i.e. the question of successful agricultural extension systems. The five chapters in this section cover major types of national agricultural extension systems, examples of extension systems actually employed by selected countries, and a discussion of the importance of linkages in making agricultural extension effective.

George H. Axinn, at the time a consultant for the Food and Agriculture Organization of the United Nations to the Government of Nepal, provides an overview of 'The Different Systems of Agricultural Extension Education'. Axinn reviews both external and internal factors contributing to the success or failure of agricultural extension efforts, with special emphasis on Asia and Africa. He makes an important distinction between 'delivery' (top-down) agricultural extension systems and 'acquisition' systems where farmers initiate and control the requests for technology transfer. He notes that an examination of the various agricultural extension system models can quickly reveal who controls their purpose and who are the intended beneficiaries.

Robert E. Evenson, professor of economics at Yale University, analyzes 'The International Agricultural Research Centers (IARCs) and their Impact on National Research and Extension Programs'. His report provides insights into the economics of extension and indicates that indeed the development of the IARC system has produced a measurable impact on the size and character of national agricultural research and extension programs. Evenson's methodology for measuring agricultural research and extension spending and the impact of the IARCs on this spending will be of particular interest to researchers. In brief, Evenson provides (1) a descriptive summary of national research and extension spending in various countries since 1959; (2) the rationale for national research and extension investment; and (3) a summary of calculations based on an econometric study of the determinants of investment in national research and extension from which he draws inferences regarding IARC impact.

Daniel Benor, consultant to The World Bank, outlines the principles of the 'Training and Visit Extension' (T&V) system, advocating a return to the basics. Claiming that key aspects of the T&V system are often misunderstood or ignored, he discusses the principles of T&V, misinterpretations of the system, and examples of current work in Burkina Faso. Benor compares and contrasts the Burkina Faso traditional agricultural extension system with the recently established T&V-derived system operating parallel to it.

Christopher A. Onyango, Chairman of the Department of Agricultural and Extension Education at Egerton College, Njoro, Kenya, applies system principles to a particular country situation in 'Making Extension Effective in Kenya'. He organizes his essay around four topics: (1) the decentralized 'District Focus' which has become a major policy and development strategy in Kenya; (2) agricultural systems structure and management as reflected in the organization of the Districts and their responsibilities, with a commentary on the T&V system; (3) incentive systems that operate in the public and private sectors; and (4) farmer participation in production targeted programs, developed through the efforts of agricultural extension as well as by the farmer organized Agricultural Society of Kenya.

Abraham Blum considers 'The Israeli Experience in Agricultural Extension and its Application to Developing Countries'. Agricultural extension in Israel is rich in experience, based on its differing programs in kibbutzim, moshavim, and in the Arab sectors. Blum comments on the main features that have made the Israeli experience successful, emphasizing that some may not be easily transferable from country to country, viz: (a) a vision of what agricultural research and extension can do to advance development; (b) dedication to that vision; and (c) inventiveness by those who are dedicated to the vision.

The third section of this volume is a compilation of five chapters which highlight selected EMERGING PRIORITIES in agricultural extension. The first three chapters speak to specific priorities and the final chapter provides a general review of emerging priorities for developing countries in agricultural extension.

Celia Jean Weidemann, formerly with the Midwest Research Institute, is currently a private consultant. Her discussion of the problems of 'Designing Agricultural Extension for Women Farmers in Developing Countries' focuses on how agricultural extension can span the gender gap and increase productivity of the significant numbers of women by-passed by traditional agricultural extension systems. She

(1) reviews the statistics on the participation of women in developing country agriculture; (2) analyzes their interaction specifically with US Agency for International Development (US/AID) projects; and (3) proposes how traditional extension models can be modified to reach women farmers and thereby increase productivity.

Jon R. Moris (at the time on the faculty of Utah State University and currently with the Overseas Development Institute (ODI) emphasizes the importance of 'Incentives for Effective Extension'. He reconceptualizes how organizational contexts can promote or inhibit agricultural extension, using illustrations from public services in East Africa. Moris considers the importance of such incentives as price ratios and credit (and input) subsidies for farmers. He then reviews incentives for agents: adequacy of technical packages, the challenge of coordinating bureaucracy 'from below', and the problem of untenable working conditions. Moris emphasizes that the main issue must be improvement at the field level. This leads to a review of the FSR approach to technology generation and the T&V structured approach to technology diffusion, as they have developed in East Africa. Moris argues that new approaches (such as FSR and T&V) represent a significant improvement over traditional agricultural extension approaches and that integrating FSR and T&V may be both feasible and practical for Africa. He recommends two major priorities: (1) that experimentation with new systems be continued; and (2) that the integration of new systems be further explored.

William M. Rivera, University of Maryland, examines 'India's Agricultural Extension Development and the Move toward Top-Level Management Training'. This essay is organized into four parts. Part one reviews the Union Government's agricultural extension support efforts at the national level and the relationship of the central government to the state-directed agricultural extension systems, almost all of which are based on the T&V system. This is followed by a discussion of the state-run near nationwide T&V extension system and its evolving management priorities. Third, a brief historical analysis is undertaken of the GOI (Government of India) move toward agricultural extension management training for top-level officials. Finally, a preliminary senior-level agricultural extension management curriculum is proposed, based on the work of a Food and Agriculture Organization (FAO) team sent to assist the GOI in March/April 1986. The chapter concludes that agricultural extension management for top-level officials is needed, but that care should be taken when organizing such training to distinguish between budget arm and implementation arm officials in state Departments of Agriculture. The chapter also suggests that regional development of agricultural extension management training institutes should be a priority for donor agencies concerned

7

with agricultural development and supporting extension systems.

Wajih D. Maalouf describes FAO's Action Plan for Agricultural Education and Training in Africa in 'Agricultural Manpower Development in Africa'. The importance in Africa of training manpower for agricultural tasks was recognized as a priority by the 12th FAO Regional Conference, and resulted in two major studies - one, a survey which produced a Directory of existing training facilities in Africa, and the second an assessment of trained manpower in the 47 countries studied. These studies reveal that while some countries have many training facilities (Egypt - 86, Nigeria - 65, Zaire - 47, Tunisia - 46) others (Botswana, Congo, Equatorial Guinea, Mauritania, Swaziland) have only one each. Also noted is that only 15 per cent of all students enrolled in agricultural training institutions were women. Indeed, of the 400,000 trained agricultural personnel in the 47 countries women represent only 3 per cent of the total. Maalouf enumerates several priorities suggested by the survey that are applicable to the African Continent as a whole.

Michael Baxter, senior adviser on rural development in The World Bank, summarizes a number of 'Emerging Priorities for Developing Countries in Agricultural Extension'. Among these are (1) the development of communications systems and techniques, which is receiving renewed emphasis as technology develops; (2) the increased attention to privatization and cost recovery, as public funds become tighter worldwide: (3) experimentation with group and individual agent/farmer contacts, underlining the concern for delivery variations; (4) improving agricultural research/extension linkages - a critical and continuing problem; and (5) serving women farmers, a need and responsibility which are ostensibly finally being recognized.

The final chapter of this publication is an epilogue which provides the editors with the opportunity to review and discuss the contributors' insights and to put forward selected observations stimulated by the colloquium series and during the compilation of this book. It is hoped that readers will find this volume to be informative and that it will prove to be a helpful tool for study and decision making in the advancement of international extension development.

REFERENCES

Axinn, G.H. and Thorat, S. (1972) Modernizing World Agriculture. New Delhi: Oxford & IBH.

Cernea, M.M., Coulter, J.K. and Russell, J.F.A. (1985) Research-Extension-Farmer: A Two-Way Continuum for Agricultural Development. Washington, DC: The World Bank.

Feder, G. and Slade, R. (1984) A Comparative Analysis of Some Aspects of the Training and Visit System of Agricultural Extension in India (Report no: ARU 19). Washington, DC: The World Bank, Agricultural and Rural Development Department.

Lowdermilk, M.K. (1985) 'A System Process for Improving the Quality of Agricultural Extension', Journal of Extension Systems, 1, 45-53.

Moris, J. (1981) Managing Induced Rural Development. Bloomington, Indiana: International Development Institute, Indiana University.

Moris, J. (1983) What Do We Know about African Agricultural Development? The Role of Extension Performance Re-Analyzed. Washington, DC: U.S. Agency for International Development, Bureau of Science and Technology.

Mosher, A.T. (1978) An Introduction to Agricultural Extension. NY: Agricultural Development Council.

Rivera, W.M. (1986) Comparative Extension: The CES, TES, T&V and FSR/D, Occasional Paper # 1. College Park, MD: The University of Maryland, Center for International Extension Development (CIED), College Park.

Rivera, W.M. (1987) Planning Adult Learning: Issues, Practices and Directions. London: Croom Helm.

Schuh, G.E. (1985) Strategic Issues in International Agriculture (draft paper). Washington, DC: The World Bank.

I. ISSUES

Chapter One

THE PRIVATE SECTOR: ITS EXTENSION SYSTEMS
AND PUBLIC/PRIVATE COORDINATION

William L. Rogers
U.S. Agency for International Development

INTRODUCTION

Public sector extension, although not without some success,
has generally been disappointing in transferring improved
agricultural technologies from research to the farmer in less
developed countries (LDCs). Extension institutions and programs exist in virtually every developed and developing
country and yet, in the latter, the coverage of farm families
is still too limited. As well, the effectiveness of government
extension systems as a viable technology diffusion method has
been seriously questioned by donor agencies. Is there justification for continuing to support and strengthen extension in
its present form? If not, can extension be reoriented, re-
directed and transformed into a more viable force for technology transfer? How might this be accomplished?
    This paper discusses alternatives to the typical LDC
government agricultural extension system. The major alternative, and the focus of this paper, is private sector extension.
What is the role of the private sector in agricultural extension, how does it function, and how is it working in selected
developing countries?

PRIVATE SECTOR ROLE

In an LDC rural setting, the agricultural private sector is
extremely diverse. Depending on the particular economic and
political situation, the private sector may consist of individual
farmers/farm enterprises of all sizes, agricultural input

---

*The views and opinions expressed in this chapter do not
necessarily reflect the position or policy of the United States
Agency for International Development and no official endorsement should be inferred.

industries, agro-service enterprises, processing industries, marketing firms, and multinational firms and/or their subsidiaries.

It also may include a wide range of agricultural production and marketing cooperatives, farmer associations and private and voluntary organizations. Despite their differences, all of these enterprises share a common market-orientation - they all try to make a profit by selling goods and services. As a result, all of these private sector organizations have a strong incentive to deliver goods and services (including agricultural extension) efficiently and effectively.

Private sector enterprises become involved in extension because they believe this involvement will increase their profits or enhance their ability to survive. Agricultural processing firms, for example, may enter into contractual agreements with groups of small and medium size farmers and/or with producer cooperatives, providing extension services and inputs as a means of assuring the supply and quality of the particular raw material or commodity for their factory. Firms that supply agricultural inputs such as seeds, chemical fertilizers, pesticides etc. may provide farmers with a wide range of technical and managerial information (through various outreach mechanisms) both to assure that their products are used correctly and also to increase agricultural production and income to the farmer. This also assures more customers to buy more products in the future.

A recent Agency for International Development (AID) study of agricultural credit, input and marketing services concluded that public, private and mixed delivery systems each have advantages in particular situations:

1. Public institutions are preferable when benefits are diffuse, public policies need changing and/or increased economic equity is a primary goal.
2. Mixed public/private entities work best when agricultural services not only require intensive, responsive and flexible management, but also need political influence to achieve program objectives.
3. Strictly private firms perform best when flexible management and direct and continuing interaction with farmers are needed.

This suggests that private sector extension does have a role in Third World agriculture and can be an important supplement to government extension systems for certain groups of producers under certain conditions. Private sector organizations can play a predominant extension role for particular inputs, particular outputs (i.e. commercial crops and commodities) and for particular farmers in particular geographic areas. Private firms cannot substitute completely for public agencies. They have less to contribute when:

1. The policy and regulatory environment is poor;
2. When target populations are remote;
3. When infrastructure is lacking; and
4. When production is mainly basic food commodities grown by subsistence farmers [1].

PRIVATE FIRMS AND THEIR ROLE IN EXTENSION

When private firms become involved in extension, mutual benefits result. By helping farmers benefit from increased incomes and economic security, firms can benefit too - by earning profits or achieving other strategic objectives. Like government extension systems, private sector extension activities vary widely from those firms which provide information on new products to those which specify complex production and management practices. It is the latter type of firm which most concerns us here. Coordination between agro-industries and small and medium sized farmers holds considerable potential for rural development if it can facilitate the transfer of technology, increase production, and thus integrate the rural population into the national economy.

More specifically, private firms become involved in extension to:

1. Promote sales of production inputs or services such as seeds, fertilizers, pesticides, herbicides, tools, machinery, animal feeds, veterinarian medicines and supplies;
2. Assure a continuous supply and/or quality of agricultural products for marketing and/or processing; and
3. Promote or protect returns on investments in farms (in the case of a bank or private developer).

The mechanism for providing this type of outreach or extension is the production management contract. This contract between the agro-industrial firm and the grower provides instruction to the farmer not only about what to produce, but how to produce it. Farmers are ordinarily not willing to accept production information without knowing its ultimate value. A marketing contract, however, guarantees its value. Similarly, the agro-industrial buyer of the raw material is not willing to provide the extension services unless it can perceive a benefit. With a contract, it can assure itself of the returns to this technical assistance by deducting the cost of the extension service from the crop price.

EXAMPLES OF PRIVATE INVOLVEMENT IN
LDC AGRICULTURE

Guatemala

In several less developed countries, agro-industrial firms provide technical advice, inputs and credit to groups of small farmers. These farmers are organized around a processing firm. Satisfactory returns are achieved within short time frames. A case in point is the operation of ALCOSA (Alimentos Congelados Monte Bello, S.A.) in Guatemala. This firm, a wholly-owned subsidiary of Hanover Brands, purchases and freezes vegetables (cauliflower, broccoli, brussel sprouts, snow peas and okra) for export to the United States. According to recent data, ALCOSA has purchased 11 million pounds of these products from 2,000 farmers, 95 per cent of them small farmers (1-6 acres - with a mean size of 2.6 acres).

The company operates buying stations to purchase cauliflower and broccoli in 17 small highland villages. They also operate three research sites. Extension consists of field employees, assigned to several villages. Field personnel contract for the product, supply inputs and technical instructions and grade and buy the product at harvest time.

ALCOSA employs a field staff of approximately 18 persons: one director of crop operations, one chief agronomist, two agronomist assistants and up to 14 local assistants. Farm production in each zone begins with a series of visits by the agronomists and their staff a month or two before the highland dry season comes to an end. In these meetings the agronomists identify the farmers who will be producing cauliflower and broccoli for ALCOSA. Arrangements are made for the supply of inputs such as seeds or transplantable seedlings, fertilizer and insecticides, as interest-free loans against the harvest deliveries. The amount of inputs needed per farmer is based on calculations made by the agronomist after the farmer has signed a contract. The farmers are not obligated to buy their inputs from ALCOSA and indeed, some choose not to do so. The latter either finance their own production or use their ALCOSA contracts as evidence of an assured market to apply for bank credit. Originally cash was paid for the product at time of purchase, but this has now been changed to weekly, and in some cases, monthly payments. One of the key inputs has been market transportation. During harvest periods, trucks bring empty baskets from the factory and carry the packed, classified and weighed baskets of product back to the processing plant. The farmer receives a receipt for his delivery and the product has thus formally been transferred to the company.

Over the years of this project, effects have included the following:

Agricultural
- Cultivation patterns have changed from diversification to concentration on the cash crop.
- The use of production credit, previously non-existent in the villages, is now common practice. Investments in inputs have risen.
- Technical knowledge and know-how have increased.
- Average investment in small equipment (horses, sprayers, containers, etc.) has increased 200-400 per cent. Investment of both inputs and labor per acre of vegetables has increased.

Economic
- Farm incomes have increased, although not as rapidly as predicted.
- Reject and overflow crops are utilized in the domestic market.

Social
- Farmers in some highland villages have organized themselves into cooperatives. Poor or lower stratum farmers have increased their incomes allowing them more independence as family farmers versus farm laborers [2].

Dominican Republic
Agro Inversiones Compania por Acciones (AI) is a fruit and vegetable enterprise located in the Azua Valley. Production is based on a satellite system of procurement, with 110 small scale farmers currently under contract to produce melons for packing and export to the United States. The primary investment of AI is in a packing plant. Twenty to twenty-five per cent of its recurrent operating cost is allocated to extension services. These services consist of five agronomists who provide technical advice to the growers. AI also supplies seed and all other inputs on credit which is recovered at the time of the delivery of the melons to the packing plant.
    The relationship between the company and the farmers is one built essentially on economic grounds. Extension interaction relates to rising incomes for both parties [3].

Mexico
Productos del Monte (PDM) is a food processor, canning 69 different fruits and vegetables for the Mexican market. Only one item, canned white asparagus, is exported. PDM, along with several other food processors in the area, pioneered the development of satellite farming in Central Mexico. By 1983 PDM was contracting 8,750 acres, held by 140 farmers, 130 of

whom were private landowners and 10 of whom were ejidatarios. Contract farmers supply 80 per cent of cannery requirements; the rest is purchased on the open market. A staff of nine agronomists is responsible for the satellite farming supply system. The farmers negotiate their credit needs with banks, but all other technical on-farm assistance, seed, spraying and machinery rentals are the responsibility of the company [4].

PRIVATE AND PUBLIC SECTOR COORDINATION

There are a number of examples of coordination between the public and private sectors in LDC agricultural development. This cooperation includes the public sector of both the US government and Third World governments. In agricultural research, extension, credit and finance and infrastructural projects, it is very common, indeed essential in many cases, that government be involved. This involvement, however, should preclude parallel or complementary private sector activity. For example, private seed and fertilizer/pesticide companies have been important research units in the US and Europe.

One public entity that has worked closely with the private sector on Third World development projects is the US Agency for International Development.

AGENCY FOR INTERNATIONAL DEVELOPMENT (AID)

AID, in its role as a development agency, cooperates with the private sector, both US and indigenous firms, to promote LDC agricultural production. Current AID policy is to support a substantial transfer of responsibility to competitive markets and private enterprises. It is trying to seek creative ways to increase the involvement of the private sector in traditional government programs such as agricultural research and extension. In particular, AID encourages the development of new arrangements between input suppliers, farmers and marketing firms such as 'contract' or 'satellite' farming [5].

An example of this type of public/private sector cooperation is an AID-funded collaborative project in Belize. Hershey Foods Corporation is cooperating with a private voluntary organization, VITA, and the Pan American Development Foundation to establish a demonstration farm for the production, fermentation and drying of cocoa. Also involved are the extension service of the government of Belize and the Peace Corps. A key component of the 3 year, $1.8 million project is the training of small family farmers and government extension service personnel in the use of improved seedlings and new production and processing methods. Hershey's

Research Farm facilities are being used for the training site, as well as for the establishment of a large nursery for improved varieties of cocoa seedlings. Other smaller nurseries are also being established to serve small groups of family farms. In three years the project hopes to have planted 500 acres of new cocoa seedlings on individual farms and trained 50 farmers in improved production process. In addition, six extension agents will have been trained to transfer the technology to farmers in other areas of the country [6].

Another AID-funded example was an experimental project in south-central Chile. This program, located in Curico, organized a group of fifty to sixty small farmers. The farms ranged in size from 9-16 hectares, averaging 11.5 hectares. Extension services received were selected and paid for by the farmers themselves at the time they received payment for harvest. The majority of the region's farmers produced onions. Root crops were also produced. The farmers hired a University-educated agronomist (Ing. Agronomo), an individual with practical experience who had previously lived and worked on one of the large farms in the district. This individual, who also operated an agricultural consulting firm, provided technical and management assistance for a fee, based on a percentage of the market value of the crop. Extension services consisted of individual farm visitations as well as group seminars. Farmers were visited once every four to six weeks, depending on the season. Discussions focused on crop varieties, sources of inputs, bank credit applications, etc. Individually and collectively, farm production rose in the first year. With few exceptions, net farm income increased over the previous year. Farmers who had previously farmed under a tenant and then collective system, now owned and farmed their own land and were able to do so on a commercially viable basis.

CONCLUSIONS

The conclusions to be drawn are not all that obvious and do not fit the usual development model of Third World extension projects. As Rivera points out, technology transfer, while at the core of extension's purpose in developing countries, is not the only purpose. Rather, extension is part of a long range rural development process.

Does the private sector have a role in extension, and thus in rural development? The answer appears to be affirmative, especially in the utilization of the core-satellite model, where corporate food processors link up with small farmers through production contracts, exchanging agricultural inputs and services for assured deliveries of produce. This model is only feasible, however, when specific economic, technical and social conditions prevail. As well, the model depends for its success on active government support.

Some literature is critical of private extension systems, especially as it is utilized in contract farming. It is viewed as a method whereby agribusiness controls agricultural production while transferring all risks to the growers. The actual situation is quite different in that growers can and do withdraw from unprofitable schemes. Contract violations by the company are seen to be a sign of a failing concern and growers quickly look for alternative markets.

In summary then, it can be concluded that under certain conditions and working with commercial, not subsistence farmers, private sector extension can be extremely effective in agricultural production and rural development, including social equity.

NOTES

1. US Agency for International Development, Stimulating Private Sector Extension. Author, Washington, DC (1985).
2. K. Kusterer et al., The Social Impact of Agribusiness - A Case Study of ALCOSA in Guatemala. US Agency for International Development Special Study No. 4. USAID, Washington, DC (1981).
3. S. Williams and R. Karen, Agribusiness and the Small-scale Farmer. Boulder: Westview Press (1985).
4. Ibid.
5. US Agency for International Development, Private Enterprise Development. Author, Washington, DC (1985).
6. P.H. Rogers, Better Coca, Better Markets: PVO's Team with Hershey to Help Farmers in Belize. VITA News (1985).

REFERENCES

Austin, J. (1981) Agroindustrial Project Analysis. Baltimore, MD: The Johns Hopkins University Press.
Daines, W.L. et al. (1979) Agribusiness and Rural Enterprise Project Analysis Manual. Washington, DC: US/AID.
Goldberg, R. (1974) Agribusiness Management for Developing Countries-Latin America. Ballinger Publishing Company.
Goldsmith, A. (1985) 'The Private Sector and Rural Development: Can Agribusiness Help the Small Farmer?' World Development.
Minot, N.W. (n.d.) 'Contract Farming and its Impact on Small Farmers in LDCs'. Paper Funded by US Agency for International Development.
Ray, H.E. (n.d.) Incorporating Communication Strategies into Agricultural Development Programs. Washington, DC:

Academy for Educational Development.

Rivera, W.M. (1986) Comparative Extension: CES, TES, T & V, FSR/D. College Park, MD: University of Maryland, Department of Agricultural and Extension Education.

Truitt, G. (n.d.) Multinationals: New Approaches to Agricultural and Rural Development Programs. Fund for Multinational Management Education and Aspen Institute for Humanistic Studies.

US Agency for International Development (1983) The Private Sector: The Dominican Republic. AID Evaluation Special Study No. 16, Author, Washington, DC.

US Agency for International Development (1985) Agricultural Credit, Input and Marketing Services, Evaluation Report No. 15. Washington, DC.

Chapter Two

ADMINISTRATIVE DECENTRALIZATION OF AGRICULTURAL
AND RURAL DEVELOPMENT PROGRAMS IN ASIA:
A COMPARATIVE ANALYSIS*

Dennis A. Rondinelli
Research Triangle Institute

Governments of developing countries in Asia and the Pacific
have been experimenting cautiously over the past decade and
a half with policies and programs that seek to decentralize
development planning and administration. The experiments are
noteworthy in part because they are unusual; these governments are highly centralized in practice if not in structure
and carry out most of their activities by central direction or
control. For a variety of reasons to be examined later, they
have chosen decentralized administrative arrangements to
implement some of their development programs, but little
attempt has been made thus far to analyze them in comparative perspective. Sporadic and preliminary evaluations of
individual programs suggest that results have been highly
variable - some were successful in achieving a few but not all
of their objectives; a few produced the desired results in
some provinces and districts but not in others; and some
failed to achieve any of their intended goals but yielded
lessons that were used to revise and improve subsequent
experiments [1].

Although these programs do not constitute the whole of
any country's experience with decentralization, they are
noteworthy because they reflect the factors that influence
governments' ability to pursue high priority objectives and to
diffuse responsibility for development beyond the central
bureaucracy. Thus, analysis of the factors that influenced
the success or failure of these new programs can contribute
to improving administration in developing countries and to
formulating and implementing programs that generate economic

---

*This is a revised version of a paper prepared for the Project
on Implementing Decentralization Policies and Programs at the
United Nations Center for Regional Development in Nagoya,
Japan. The opinions and conclusions are those of the author,
however, and do not necessarily reflect those of the UNCRD.

growth with greater social equity, an avowed aim of all of the governments that initiated them.

This paper summarizes and analyzes experience with decentralized development in Asian and Pacific countries through nine case studies commissioned by the United Nations Center for Regional Development [2]. From a review of these cases, the rationale and purposes of decentralization are discerned; the forms of decentralization that were used and their general impacts and results are described; and the administrative, political, behavioral, economic and physical factors that influenced their implementation are analyzed.

APPROACHES TO DECENTRALIZATION

The variety of ways in which governments have attempted to decentralize development planning and administration is shown in Table 2.1. Indonesia, the Philippines, Thailand and Fiji used provincial administrative units; Pakistan and Nepal created integrated rural development programs in which responsibility was deconcentrated to regional or district organizations; Malaysia and India used semi-autonomous authorities to pursue agricultural and rural development; and Sri Lanka strengthened district administration.

It is important to note that in all of these cases the initiative for decentralization came from the central government and not from demands for participation or devolution of authority from below. In every case funding for the experiments also came primarily from central revenues, although in the Philippines, Nepal and Indonesia the United States Agency for International Development (USAID) helped to initiate and finance them and provided technical assistance in their implementation. In Pakistan several international agencies contributed funds and advice. In Thailand, Pakistan, the Philippines, Indonesia and Sri Lanka a central government ministry or agency played an important role in guiding or supervising the programs. Even in those countries where a province, district, or special authority was given responsibility for development activities, the central government exercised pervasive influence. Thus, in a real sense, attempts at decentralization in Asia and the Pacific have been national policies, and analysis of them can provide insights into the variables affecting policy implementation in developing countries.

THE RATIONALE FOR DECENTRALIZATION

The nine cases reveal a wide range of reasons for decentralizing development planning and administration in Asia. Among them five seem to dominate:

Table. 2.1: Decentralization case studies

Country	India	Thailand	Malaysia	Pakistan	Indonesia
Program	Small Farmers' Development Agency (SFDA)	Rural Employment Generation Program	Federal Agricultural Marketing Program	Integrated Rural Development Program	Provincial Development Program
Case Study Location	Alwar District, Rajasthan	Lampang Province	Kuala Selangor District	Mananwala and Harappa Marakaz, Punjab Province	Madura Province, East Java
Level of Decentralization	District	Province/Tambon	Special Authority	Markaz	Province/District
Form of Decentralization	Deconcentration/ Delegation	Deconcentration/ Deconcentration	Delegation	Deconcentration	Deconcentration
Type of Program	Agricultural/ rural development	Community Facilities, small-scale infrastructure	Agriculture production and marketing	Agricultural development and related services	Small scale rural development/small-scale credit
Lead Agency	SFDA	Prime Minister's Office	Federal Agricultural Marketing Authority	Ministry of Local Government and Rural Development	Provincial Administration/Districts
Source of Funding	Central Government	Central Government Grants	Federal and State Governments	Central Government and International Agencies	Central Government and USAID
Period of	1970-1982	1975-1982	1975-1981	1972-1982	1977-1982

Table 2.1: Decentralization case studies (continued)

Country	Nepal	Philippines	Sri Lanka	Fiji
Program	Rapti Valley Integrated Rural Development Program	Provincial Development Assistance Program	District Development Planning and Management	Provincial Councils
Case Study Location	Dang District Rapti Zone	Pangasinan	Country-wide	Country-wide
Level of Decentralization	Region/District	Province	District	Province
Form of Decentralization	Deconcentration	Deconcentration	Deconcentration/ Devolution	Devolution
Type of Program	Agricultural/Rural Development	Agricultural, Rural Infrastructure, Local Fiscal Administration	Rural Development	General Local Adminstration
Lead Agency	Rapti Integrated Project Coordination Office	Ministry of Local Government and Community Development/ Province Government	District Development Council	Provincial Councils
Source of Funding	Central Government and USAID	Central Government and USAID	Central Government	Central Government Grants and Pro-vincial Revenues
Period of Case Study	1977–1982	1968–1982	1965–1982	1967–1982

1. In many countries decentralization policies were adopted because of the disappointing results of our recognized deficiencies in central planning and management. The limits of central planning in directing development at the local level became evident during the 1970s, as did the inflexibility and unresponsiveness of central bureaucracies in many countries. Decentralization was seen as a way of overcoming or avoiding these constraints [3].

2. During the 1970s and early 1980s the emphasis of development policies changed in many countries away from maximizing economic growth and toward promoting more equitable distribution of the benefits of development, reducing disparities in income and wealth between urban and rural areas and among regions, and increasing the productivity and income of the poor. Equitable growth policies strongly implied the need for programs that were tailored to local conditions, that elicited the support and involvement of local administrators and of the people they were intended to help, and that integrated the variety of services required to stimulate the economies of rural areas [4].

3. The growing involvement of government in promoting widespread, non-traditional development activities made it clear during the 1970s that complex and multifaceted programs were difficult to direct and control exclusively from the center. Decentralizing development planning and building the administrative capacity of local organizations seemed essential to improving the effectiveness of the central government, as well as the ability of local administrative units to deliver services needed for development, especially in poor and remote rural regions.

4. Crises or external pressures to act expeditiously in some countries highlighted the difficulties and constraints of working through entrenched central bureaucracies and forced national leaders to search for alternative ways of coordinating activities at the provincial or local levels to solve serious social, political or economic problems quickly.

5. Decentralization in some countries was associated ideologically with principles of local self-reliance, participation, and accountability and was pursued as a desirable political objective in itself.

The cases reveal that it was a combination of these factors that usually led central government officials to propose decentralization and that often the rhetoric by which the proposals were justified veiled other motives and intentions. Central government officials in many countries believed that the new arrangements would elicit support and cooperation from local communities for national development policies, and decentralization was often viewed as an instrument for extending the central government's influence or control.

The rationale for decentralization varied among countries. In India, the Small Farmers Development Agency was established partly in reaction to the failure of panchayati raj and other experiments in local democracy. Since neither the central bureaucracy nor local governments could be relied upon to deliver services and to involve local residents in decision making, other arrangements had to be found to increase agricultural production quickly when India faced a severe food crisis in the 1960s. With encouragement from international organizations, the government established special project units to integrate its services locally. But neither the Intensive Agricultural Development Program (IADP) nor the Integrated Rural Development Program, which did increase agricultural output during the 1970s, were effective in raising the productivity or incomes of small scale cultivators and landless laborers. The All India Credit Review Committee of the Reserve Bank of India was concerned that the Green Revolution would lead to more serious disparities than already existed between large and small scale cultivators, unless ways could be found quickly to provide the latter with credit, technical assistance and support services to increase their productivity and income [5]. Semi-autonomous organizations - Small Farmers Development Agencies - were created because past experience made it clear that local institutions were not strong enough to carry out participative development activities on their own and that the national civil service was not responsive or flexible enough to plan and implement such a program [6].

A similar combination of factors led to the creation of the Integrated Rural Development Program in Pakistan in 1972. The Green Revolution was also successful there in increasing agricultural production but, as in India, it exacerbated social and economic disparities in rural areas. The IRDP would sustain agricultural production and extend the benefits of the new methods and technology to the rural poor. By concentrating the local offices of national departments dealing with agricultural development, representatives of credit agencies and sales outlets for agricultural inputs in the markaz, and by coordinating their activities through a Project Manager, the government hoped to make the central bureaucracy more accessible and responsive to small farmers. The activities of central departments would be coordinated with those of farmers associations and private investors to give district residents a greater voice in development planning and administration, or at least to promote better communication between farmers and the government's agricultural service departments [7].

Underlying these programs in India and Pakistan was a long-standing ideological commitment to decentralization and popular participation in development. Khan points out that successive national plans in Pakistan emphasized these principles. The First Five-Year Plan in 1957 noted that:

> Planning in a free society must be based on a general consciousness of social purpose so that the people treat the plan as their own, intended for their benefit. They should have a sense of participation and be willing to extend their full support and cooperation in its fulfilment. Without the wholehearted participation of the people, the development program will not achieve its full proportions; progress will be slow; and its benefits will remain open to questions [8].

A major objective of the Fourth Plan issued in 1970 was to promote 'the maximum decentralization of responsibility and authority in all areas bearing upon plan implementation' [9]. But as will be seen later, this conviction was not always widely shared in either India or Pakistan and the purposes and processes of decentralization were often interpreted differently by both central government officials and local elites.

Changes in the thrust of development policies in most Asian countries during the 1970s also had a profound effect on administrative procedures. In Indonesia the creation of the Provincial Development Program was due to a 'shift of development objectives from those fostering economic growth to others promoting distributional equity and widening popular participation in development planning and implementation' [10]. The U.S. Agency for International Development provided funds and technical assistance in 1977 to decentralize planning to the provinces and to help the central government increase the income and improve the living conditions of low income villagers. This was to be done by expanding the administrative capacity of central ministry field officers and local government officials. Regional Planning Agencies (BAPPEDAs) were to extend project planning, implementation and evaluation capacity to the provinces because the small scale rural development projects envisioned by the Indonesian government could not be effectively designed and carried out by the central agencies alone.

Increasing regional disparities and rural poverty in Thailand during the late 1960s and early 1970s also led the central government to search for more effective ways of raising the income of the rural poor. The Rural Employment Generation Program sought to provide paid jobs for the poor during the dry season when farmers and farm workers were either idle or migrated to the cities looking for temporary employment. Recognizing that central ministries could not identify and formulate the myriad small scale projects needed to absorb rural labor, political leaders deconcentrated the program to provincial governments, which in turn sought to develop the capability of Tambon Councils to implement community development activities. In this way, the government could provide employment and also increase support for national policies among rural people, especially in areas where poverty led to social unrest and external subversion [11].

Attempts at decentralizing development administration in Sri Lanka were also closely related to the basic principles of national development, which included promoting a geographically widespread distribution of development activities and the benefits of growth; redistributing income to poorer groups within society; generating greater employment opportunities; creating self-sufficiency in food production; and providing social services and facilities to a large majority of the population. There was a strong belief in Sri Lanka that achieving these goals required popular participation. But participation, as Wanasinghe notes, has been perceived of 'more as an instrument mobilizing support of the public for specific projects and activities rather than as a state wherein the public participate directly in the decision making process' [12]. The central government did not interpret its advocacy of popular involvement to require devolution of planning and implementation responsibilities. However, it was the worsening food crisis of the mid-1960s and the need to increase agricultural production among small scale farmers that created strong dissatisfaction with central administration and the fragmentation of development programs at the district level. At the village, divisional and district levels the national departments of Cooperative Development, Agricultural Marketing, Agrarian Services, Agriculture and Irrigation were each pursuing their programs in isolation and responding to directions from their own headquarters in Colombo, without regard for the impact of each others' actions on rural communities. The senior public official in the district, the Government Agent, had little control over the decisions or operations of national departments within his jurisdiction [13]. When it became clear that such arrangements would not achieve national goals and that district residents would not become enthusiastically involved in development activities, a different concept of decentralization and participation began to emerge. By 1980, the Presidential Commission on Development Councils would argue that:

> Economic development as we see it is not a mere matter of 'growth' measured in terms of percentages of Gross National Product. We prefer a judicious blend between growth as such and ... 'the quality of life'. That is not

a mere material concept. It is more comprehensive than this in the sense that the totality of the environment in which it takes place is just as important as economic growth itself. This shift in emphasis has other implications. It takes us logically in the direction of decentralized administration. Economic development is a mere exercise in bureaucracy if the people of the localities in which it takes place and whom it is intended to benefit do not share in the responsibilities of decision-making [14].

Thus, responsibility for local development was devolved to District Development Councils in 1980.

In Fiji, three factors explain the devolution of local government functions to Provincial Councils. As in the other countries, decentralization in Fiji was adopted in association with a strategy for promoting more equitable development. The Seventh Plan for 1976 to 1980 called for increased emphasis on regional development to alleviate growing disparities in income and for actions to 'decentralize economic activity by location and broaden involvement by race and enhance opportunities, material living standards and social and cultural amenities of the rural areas' [15]. Drastic administrative reforms introduced in 1967 attempted to eliminate vestiges of colonial rule and to create a unified national government. One of their major objectives was 'to extend the authority of the central government over all racial groups and to bring the whole country under one uniform administration' [16]. Creation of Provincial Councils as local governments with a constrained scope of authority must be understood in the context of those attempts to strengthen central governance. Later decentralization was justified by the rapidly expanding role of the government in political, economic and social affairs and its inability to control those activities from the center.

Malaysian leaders turned to such public corporations, as the Federal Agricultural Marketing Authority to avoid the constraints of the regular bureaucracy in pursuing high-priority political and economic objectives. With a greater degree of independence special authorities could 'undertake their jobs with a greater sense of urgency and purpose ... and be free in developing their own core of trained personnel and employing them on their own schemes of service' [17]. In the case of FAMA, delegation was necessary because the government could not regulate the market directly to increase the access of poor Malay farmers. The private sector was dominated by Chinese traders, who were not organized to collect and sell the output of numerous and widely scattered small holders and padi planters, nor were they particularly interested in doing so. Delegating these functions to a public corporation was the only real alternative open to the government.

The Provincial Development Assistance Project (PDAP) in the Philippines was based on the assumption that decentralization could not be meaningful unless local governments had the technical and managerial capability to plan and implement local development activities. 'In specific terms', Iglesias points out, 'lack of local capability actually translates into shortages of trained manpower, lack of financial resources, and the fact that the major development tasks of local development were the responsibility of agents of national ministries' [18]. The rationale for PDAP was that development of local technical, managerial and financial capacity must proceed or be undertaken concomitantly with decentralization.

In Nepal, USAID supported the decentralization of integrated rural development in the Rapti Zone because both aid agency and central government officials recognized that the severe problems of poverty in rural areas could not be solved from Kathmandu, which had weak communication and transportation linkages with remote regions. Moreover, the central government had limited financial, managerial and other resources with which to cope with problems of widespread rural poverty. If the problems were to be solved, the capacity of local officials to identify, plan, finance, and carry out projects would have to be strengthened [19].

FORMS OF DECENTRALIZATION

Decentralization can be broadly defined as the transfer of planning, decision making or management functions from the central government and its agencies to field organizations, subordinate units of government, semi-autonomous public corporations, area-wide or regional development organizations, specialized functional authorities or non-government organizations [20]. Four forms of decentralization can be distinguished by the degree of authority and power, or the scope of functions, which the government of a sovereign state transfers to or shares with other organizations within its jurisdiction.

1. Deconcentration involves the transfer of functions within the central government hierarchy through the shifting of workload from central ministries to field officers, the creation of field agencies, or the shifting of responsibility to local administrative units that are part of the central government structure.

2. Delegation involves the transfer of functions to regional or functional development authorities, parastatal organizations, or special project implementation units that often operate free of central government regulations concerning personnel recruitment, contracting, budget-

ing, procurement and other matters, and that act as an agent for the state in performing prescribed functions with the ultimate responsibility for them remaining with the central government.

3. Devolution involves the transfer of functions or decision making authority to legally incorporated local governments, such as states, provinces, districts or municipalities.

4. Transfer to Non-government Institutions involves shifting responsibilities for activities from the public sector to private or quasi-public organizations that are not part of the government structure.

The nine cases reviewed here indicate that all four types of decentralization were used in Asia. The use of non-government institutions such as voluntary and religious groups, private enterprises, farmers associations, rural cooperatives, and others, is common throughout Asia and the Pacific, although none of these cases focused exclusively on them. Five of the programs - in Thailand, Pakistan, Indonesia, Nepal and the Philippines - illustrate the deconcentration of development planning and administration functions to subordinate units of the central government - regional, provincial and district organizations - that were financed, supervised, and monitored, if not directly controlled, by a central ministry or agency. Attempts to strengthen district planning and management in Sri Lanka prior to 1980 were also an example of deconcentration. Each case illustrates a somewhat different arrangement for deconcentrating functions and a different pattern of interaction among central, subordinate and non-government organizations.

The programs in India and Malaysia involved the delegation of functions to a semi-autonomous agency. The Small Farmers Development Agency in India combined some elements of delegation with deconcentration. Although SFDA was established as a corporate body with its own governing board, it remained under the supervision of, and was financed from grants by, the central government. Mathur prefers to call the SFDA an experiment in 'controlled decentralization' [21]. The Federal Agricultural Development Authority in Malaysia is a more conventional example of delegating functions to a public corporation, although it will be seen later that the Ministry of Agriculture maintained a good deal of indirect influence over FAMA's activities.

The transfer of authority to incorporated local governments is seen in the creation of Provincial Councils in Fiji, which have popularly elected members and the power to raise and spend revenue, and in the establishment in Sri Lanka in 1980 of District Development Councils, with elected members, revenue raising and spending powers and the authority to make their own by-laws.

Deconcentration

The UNCRD cases illustrate five variations of deconcentration. In Thailand responsibility was shifted to provinces and tambons through financial grants from the central government; in Pakistan deconcentration consisted of coordinating arrangements at the subdistrict level; in Indonesia and the Philippines planning and some administrative functions were shared by national ministries with Provinces; in Nepal a regional project coordinating office became the vehicle for expanding the capacity of district administrators to plan and implement projects; and in Sri Lanka national development activities were coordinated by district administrative committees, which were subordinate units of the central government until 1980.

1. Deconcentration through Financial Grants - Thailand

In order to generate rural development and increase the income of poor farmers, the government of Thailand in 1975 began to set aside a prescribed amount each year from the national budget to finance small scale projects in 5,000 Tambons (villages) and some low-income sanitary districts. The Tambon Council could select projects from categories determined by the National Committee on Rural Employment Generation. Most of the projects involved construction or improvement of water supply for domestic and agricultural use and construction or repair of community facilities that could increase agricultural production and household income or improve public health. The National Committee was headed by the Prime Minister and composed of high-level officials of national ministries and agencies concerned with rural development. The projects chosen by the Tambon Councils had to provide employment in the farming off-season for villagers and be reviewed and approved by provincial and district committees. The Provincial Committee was headed by the Governor and composed of representatives of national ministries working in the province. It received funds according to criteria established by the national committee - usually based on the extent of damage done by the previous years' drought and the province's need for additional income - and allocated them among Tambons, usually on the basis of their agricultural area and farm population. It also approved project proposals and suggested changes in them, fixed local wage rates and prices of materials, monitored standards of construction, coordinated activities undertaken by more than one Tambon, and supervised the implementation of the Program in the Province. The activities of the Provincial Committee were to be linked with those of Tambon Councils through District Rural Employment Generation Program Committees, which were chaired by the chief officer and composed of district officials. The district committees were responsible for reviewing and revising specifications and cost estimates for projects proposed by Tambon Councils, coordinating multi-Tambon projects, disbursing funds and supervising and monitoring the implementation of the rural employment program in the district. Both the provinces and the Tambon Councils were required to prepare annual plans and to coordinate their projects with those of the central government and local administrative units [22]. In this way the central government maintained control over the program but deconcentrated responsibility to provinces and Tambons for allocating funds according to prescribed formulae and for identifying, designing and carrying out local projects according to national guidelines.

2. Deconcentration through Local Coordination - Pakistan
In Pakistan, the attempt to improve agricultural production and increase the income of small scale farmers was made through the Integrated Rural Development Program (IRDP). The integration and coordination of resources and activities of national agencies at the local level would be promoted by designating project centers in rural towns (markaz) that could serve 50 to 60 villages over a 200 to 300 square mile area. The programs of national agencies would be integrated by concentrating their office in the markaz and promoting, through village associations, projects that would provide agricultural inputs, credit, extension and related services [23]. Village IRDP Committees and a markaz committee - consisting of representatives of each government agency and the Project Manager - were established, and the markaz committee was to work with farm cooperative federations and representatives of private businesses to integrate their agricultural development activities. The Project Manager, who was designated as secretary of the federation of village farm cooperatives and advisor to government agency representatives, would be the primary link between these two groups and the private sector [24]. He was appointed by the Provincial Department of Local Government and Rural Development but had no official authority over the field representatives of the national departments.

In the short term, the IRDP was to establish farmers' cooperatives in each village and a federation of these associations at the markaz, prepare labor-intensive and multiple crop production plans, establish model farms in the villages and the markaz, supply agricultural inputs, credit, storage and marketing facilities, train farmers and do applied agricultural research. The long run task of the Project Manager and his staff was to help transform farmers into service and production cooperatives and eventually into social cooperative farms. The IRDP staff would arrange training programs for farmers and credit on the basis of production plans, encourage cottage industry, generate off-farm employment opportunities, and help develop agro-processing, agricultural service enterprises and small agro-industries in the markaz. In addition, they were to assist farmers to mobilize savings, and plan and carry out rural public works and small scale community and infrastructure projects [25].

In 1980, with the reinstatement of elected Union Councils, the IRDP was reorganized. A Markaz Council was created, consisting of the chairmen of all Union Councils, the Project Manager, representatives of national government agencies and a district councillor. The chairman of the Markaz Council was elected from among the non-official members, and the Project Manager was made council secretary. IRDP and the People's Work Program were merged into the Department of Local Government and Rural Development, an assistant director of which served as secretary of the District Council and supervised the markaz Project Managers within the district. Projects were prepared by Union Councils, reviewed by the Markaz Council and approved by the District Council, which allocated funds to local projects [26].

3.   Decentralization through District Administration - Sri Lanka

In Sri Lanka there have been three phases in the evolution of decentralization policies. From 1965 to 1970 attempts were made to coordinate the activities of national government agencies involved in agricultural development at the district level. During the 1970s the government sought to strengthen the political influence of district coordinators by creating a district political authority and to provide financial resources for local development by creating a district development budget. In 1980, administrative reforms moved toward devolution and local government [27].

Until 1980, however, decentralization in Sri Lanka was perceived of as local coordination of national functions. Dissatisfaction with the fragmentation of and lack of cooperation among national departments led political leaders in 1965 to designate the Government Agent within each district as the coordinator of all national departments involved in agricultural development. He was designated as 'deputy head of department', providing him - at least theoretically - with administrative control over national technical officers working within his jurisdiction. Coordinating committees were established under his auspices to promote cooperation within the district and a Cabinet Committee chaired by the Prime Minister coordinated national agencies at the center.

The immediate objective of deconcentrating the coordination of agricultural development activities in the 1960s was to overcome the serious food crisis in Sri Lanka. The Coordinating Committees were charged with formulating annual integrated agricultural programs for districts and divisions; monitoring progress, overseeing the supply of support services, agricultural inputs and marketing arrangements needed to increase food production, and cutting through the bureaucratic maze at the center to solve implementation problems [28].

But for reasons to be discussed later, the Government Agents were not entirely effective. The succession of a new political party to power in 1970 led to decentralization of all development planning, implementation, monitoring and evaluation responsibilities to the district. Government Agents would be responsible for formulating and implementing development plans, again primarily by coordinating national department representatives in the district. District Development Committees were established with the Government Agents as principal officers, members of parliament from the district, and senior representatives of government departments as members. Planning officers were to assist the Government Agent in carrying out development tasks. Divisional Development Councils were to serve as the link between the district body and community organizations, cooperatives and village committees.

In addition, a member of parliament from the ruling party was appointed as a District Political Authority (later called a District Minister) by the Prime Minister to expedite action, cut through red tape, and help overcome obstacles to the implementation of district plans. The Government Agent would thus have a channel of political influence in the capital and an additional source of authority for coordinating the representatives of national agencies. The government established a district development budget in 1974 that earmarked a specific amount of money each year - based on the number of parliamentary electorates within the district - for small scale development activities selected by the district development committees [29].

4.   Deconcentration through Provincial Development Planning - Indonesia and the Philippines

Decentralization in Indonesia and the Philippines was pursued through the deconcentration of rural development functions to provincial administrative units. In both cases, provincial development programs were established by national planning and development agencies with assistance from the U.S. Agency for International Development.

In the Philippines, the Provincial Development Assistance Project (PDAP) was initiated in 1968 jointly by USAID and the

National Economic and Development Authority (NEDA) to strengthen the capability of the provincial governments to identify, design and implement local development projects in agricultural production and marketing, rural infrastructure and local fiscal administration. PDAP provided assistance to the provinces, first on a pilot basis and later throughout the country. Participating provinces had to create a Provincial Development Staff (PDS) to provide technical and management assistance to the Governor in planning and coordinating PDAP projects. USAID provided funding for staff training and technical assistance was provided only in those provinces in which the Governor was willing to exercise leadership and commit local resources to the projects. The Ministry of Local Government and Community Development supervised the program and provided guidelines for project selection, planning and management [30].

A similar program was set up in Indonesia in 1977. The Provincial Development Program (PDP) allocated funds from the central government to support small scale projects designed to raise the incomes of poor villagers, create employment opportunities or support local development activities. Policy was made by a Foreign Aid Steering Committee composed of high level officials of the National Planning Agency (BAPPENAS) and the Departments of Home Affairs, Finance, Public Works and Agriculture. The program was implemented by the Department of Home Affairs through the Directorate General of Regional Development, which reviewed all PDP proposals submitted by provincial governments before they were sent to the National Planning Agency [31].

Responsibility for planning, implementing and supervising the use of funds within national guidelines was deconcentrated to the provincial governors, who were assisted by the Regional Planning Agencies (BAPPEDAS), provincial government units, central government field officers and other provincial agencies. District heads (BUPATI), along with district administrative and technical personnel and representatives of central government departments were given responsibility for project identification, planning and execution. Provincial and district planning units were expected to formulate a development program aimed at alleviating rural poverty through processes of participative decision making and addressing the specific conditions and needs of the district [32].

5. Deconcentration through Regional Coordination - Nepal
Finally, in Nepal responsibility for development planning and administration was deconcentrated to regional development projects. The Rapti Integrated Rural Development Project, for example, began with assistance from USAID in 1977. Its objectives were to improve food production and consumption, increase income-generating opportunities for poor farmers,

landless laborers and women, strengthen the ability of panchayat administration and other organizations to plan, implement and sustain local development activities and to increase the availability and use of national social and productive services in the region [33].

The integrated rural development project in Rapti was implemented by a Project Coordination Office under the supervision of the Ministry of Local Development. The PCO was to coordinate all of the national agencies working within the district, provide assistance to villagers in adopting appropriate technology and review the feasibility of projects proposed by district agencies. It was also to provide technical assistance and training, information to national line agencies in the district for plan formulation, and evaluate the performance of on-going projects [34].

Delegation

In two countries - India and Malaysia - development functions were delegated to semi-autonomous agencies. In India, the Small Farmers Development Agency illustrated a highly controlled form of delegation, in which the incorporated bodies were supervised closely by and were financially dependent on the central government. The Federal Agricultural Marketing Authority (FAMA) in Malaysia was a more conventional public corporation to which the government delegated the functions that could not be easily carried out by regular bureaucratic agencies.

The Small Farmers Development Agency was established in 1970 to identify small scale and marginal farmers and agricultural laborers who required financial and technical assistance, to draw up plans for agricultural investments, and to help solve the production problems and improve the economic conditions of these groups. SFDA was also to help formulate and implement local projects, review the impact of proposed investments on small scale agriculturalists, assist poor farmers in obtaining adequate credit to improve production, and provide risk coverage for their loans. In addition, it was authorized to give grants and subsidies to credit institutions that helped small scale farmers, provide technical and financial assistance to farmers in improving agricultural and livestock raising practices, and strengthen farm marketing and processing organizations [35].

SFDAs were incorporated in the districts as registered societies with their own governing boards, but were also linked closely to the regular administrative structure. The governing boards were headed by the District Collector; a senior civil servant acted as project officer and three assistants were appointed by the departments of agriculture, cooperatives and animal husbandry. Members of the board represented government departments operating in the district, local cooperatives, and banks. Funding for the SFDA projects came from central government grants and professional staff were seconded from State Governments. However, SFDAs operated by their own rules and procedures, had their own offices and their employees were not part of the Indian civil service.

The government's objectives in establishing semi-autonomous corporations were to transfer central funds to the districts without routing them through the States and to create district agencies that were insulated from local political pressures. Incorporation would also allow the agencies to retain unexpended funds at the end of each fiscal year [36].

Promoting agricultural marketing among small scale farmers in Malaysia was delegated to FAMA because the government could not control the market directly and because the private sector was not organized to serve this group. FAMA was given the tasks of establishing marketing facilities and processing and grading centers, promoting new or expanded domestic and foreign markets for agricultural products, and purchasing from and selling the products of poor farmers who had difficulty marketing their goods. FAMA was also authorized to do marketing research, disseminate marketing information to farmers and regulate the practices of market intermediaries. Created as a special authority in 1975, FAMA's board of directors was appointed by the Ministry of Agriculture and its director-general was a federal civil service officer on secondment for two years. Its employees, however, were not part of the national civil service, rather they were subject to FAMA's own personnel and promotion system and were not transferrable to other government agencies. FAMA had offices in the States, but its State officers were responsible only for routine operations; all policies were set by headquarters in Kuala Lumpur.

Devolution

Finally, among the nine cases were two examples of devolution: the creation of District Development Councils in Sri Lanka in 1980 and the strengthening of Provincial Councils in Fiji.

In Sri Lanka, dissatisfaction with coordination by district development committees led in 1980 to legislation creating District Development Councils, which were to be composed of members of parliament in the district and other citizens who were popularly elected for four-year terms. The Councils were empowered to raise revenue from a variety of sources and obtain loans that would become part of the district development fund. The executive committee of the Council, composed of the District Minister, the Council chairman and two council members appointed by the District Minister with the concurrence of the council chairman, was to be responsible for formulating an annual development plan. The Council could formulate, finance, and implement projects and discharge local government functions previously performed by Village and Town Councils [38].

Changes were made in Fiji's administrative system in 1967. The powers of the colonial Fijian Affairs Board were transferred to the central government and the supervision and direction of local affairs were vested in Provincial Councils. For the first time, members of the Councils were to be popularly elected and they were granted powers to make and enforce local laws and to raise revenues through land taxes and other sources. The Councils were to serve as local governments for the native Fijian population, but not other ethnic or racial groups.

The Provincial Councils were to be a link between the central government and the villages, coordinate the rural development activities of central agencies, and articulate and help to meet the needs of rural Fijians [39].

THE EFFECTIVENESS OF DECENTRALIZED DEVELOPMENT PLANNING AND ADMINISTRATION

None of the UNCRD-commissioned cases were meant to be exhaustive evaluations; they were intended to examine the structure, process and operation of decentralized arrangements for development in particular provinces and district programs. The assessments were based on observations in specific areas that were not necessarily representative of conditions in the rest of the country.

Yet, it became clear that decentralized arrangements for subnational development planning and administration almost everywhere produced mixed results. In most cases, decentralized arrangements were successful in bringing greater attention to the conditions and needs of rural areas and additional funding for small scale development projects. Some elicited greater participation by local officials and nongovernment organizations than centrally administered programs. But all faced administrative problems, especially in coordinating the activities of national departments and agencies at the local level, in acquiring sufficient numbers of trained planners and managers, and in obtaining or being able to spend financial resources.

The Small Farmers Development Agency in India, for example, seemed to make a noticeable impact in providing services and inputs that required only distribution to individual farmers. In Alwar, the program was more instrumental in increasing the number of tubewells for irrigation, pumping equipment and cattle than in constructing physical infrastructure, providing technical assistance or strengthening local institutions [40]. In most years, however, the SFDA in Alwar was able to allocate less than half of the funds provided to it. In five fiscal years it was able to spend less than 25 per cent of its allocations, and the average for the other five years was about 73 per cent. SFDA in Alwar also suffered from rapid turnover of staff, unwillingness of local officials to innovate or to deal with local problems creatively, difficulty in translating central government guidelines into meaningful local development activities, and was highly dependent on the central government for funds and direction.

Chakrit Noranitipadungkarn found Thailand's Rural Employment Generation Program to be very successful in Lampang Province, but points out that not all other provinces did as well [41]. Lampang was awarded two of the three first prizes in the Northern Region in 1980 for successful dam and concrete bridge projects. In 1980 and 1981, REG projects in Lampang employed about 50,000 man/days of labor and provided work for from 4,000 to 5,000 people each year, increasing the income of 2,000 to 3,000 rural households. Moreover, the program helped train local officials to plan, design and carry out projects that were chosen in open meetings of Tambon Councils, whereas before, projects were selected by higher level government officials and carried out entirely by contractors, with little or no participation by villagers. The Rural Employment Generation Program set in motion a process through which Tambon leaders 'are learning how to conceive useful projects, how to get things done, how to mobilize people, how to communicate with officials and businessmen'. Perhaps most important of all, Chakrit notes, is that local leaders learned 'how to work democratically through the whole process of implementation' [42].

Yet, even in Lampang Province, there were some Tambon leaders who did not commit the time and energy needed to make the program successful, and who attempted to select projects without the participation of villagers. In Tambons that chose larger scale, more complex projects there was a shortage of skilled technicians needed to design and implement them.

In Punjab's Manawala Markaz, Khan found Pakistan's IRDP to be highly successful, and in Harrappa Markaz, slightly better than average. He notes that in both case study areas the Markaz Councils were active in undertaking local development projects, the number and value of which increased over the years [43]. Provincial governments provided much of the financing for local projects, but in each markaz people made contributions and the projects seemed to benefit a large segment of the rural population. In these areas the Markaz Councils were more successful in getting national ministries and agencies to provide infrastructure and social services, however, than they were in coordinating inputs for agricultural development. Indeed, throughout the

Punjab local representatives of national ministries lacked adequate resources to coordinate their activities and received little support from their headquarters to do so. The technical assistance that the central ministries were able to offer was often inappropriate for small scale projects. Thus, much of the success of the program in these two areas was attributable to the project managers' ability to get local leaders to work together on self-help projects for which guidelines were clear, funds were readily available, and which did not necessarily depend on cooperation from national departments [44].

Similarly, despite the fact that the Rapti IRDP project fell behind schedule and did not entirely meet its objectives, it had a noticeable impact on this poverty-stricken region of Nepal. It promoted projects that provided sorely needed drinking water for villages, trained at least 300 people in establishing and operating cottage industries, reforested 500 hectares and provided seed and chemical fertilizers for 900 hectares of land, created a small-farm credit program, employed local labor in road improvement and trained district administrators in various aspects of small scale project planning and implementation [45].

In Pangasinan Province in the Philippines, the Provincial Development Assistance Project was highly successful in increasing the number of roads, bridges, drinking water systems and artesian wells, but fell short in building the technical and managerial capability of the Provincial Development Staff and of other local officials in planning, financing and implementing projects on their own. The program in Pangasinan had a high rate of turnover among staff, which made the institutionalization of managerial and technical skills difficult [46].

Those cases in which decentralization aimed primarily at coordinating central government activities at the provincial or district levels seem to have been the least successful. The early attempts in Sri Lanka to use Government Agents to coordinate agricultural development in the districts failed; they allowed local political elites to 'consolidate their exercise of power and patronage rather than foster self-management by the people' [47]. But Wanasinghe argues that they did demonstrate the need for more effective coordinating mechanisms and focused the attention of the Prime Minister, Parliament, and the central bureaucracy on the district as a 'development locale', within which sectoral activities had to be better integrated if they were to have a greater impact. Moreover, they paved the way for strengthening the role of Government Agents during the 1970s.

The District Development Committees and district budgets that were established in Sri Lanka during the 1970s also fell short of their goals [48]. Much of the funding available through district budgets was used for infrastructure and

services rather than for productive activities and the intervention of members of parliament and local political leaders
undermined real involvement of district residents in the
development process. But these reforms highlighted the need
for devolution of development function to the districts, a form
of which came about in 1980 [49].

In Fiji, Provincial Councils have not been able to
exercise their revenue raising powers effectively, and because
of strong communal traditions can only spend their funds for
projects that clearly benefit the entire community. They have
thus been able to perform only traditional local government
functions - regulation of health and sanitation, road repair
and the like. In some cases they have not even been able to
enforce public health and sanitation rules, and, as a result,
local government institutions, including Provincial Councils,
have remained outside the main stream of development activities [50]. Rural development programs have been implemented
directly by central agencies or through village councils [51].

Thus, decentralized programs in Asia and the Pacific
have succeeded in achieving some of their objectives in some
places, but even where they have been successful serious
administrative problems have arisen, and the factors
influencing implementation must be analyzed in more detail.

FACTORS INFLUENCING IMPLEMENTATION

The nine cases commissioned by UNCRD reveal a variety of
factors that influenced implementation, among th most important of which were: (1) the strength of central political and
administrative support; (2) behavioral, attitudinal and
cultural influences; (3) organizational factors; and (4) the
adequacy and appropriateness of local financial, human and
physical resources.

Political and Administrative Support
The degree to which national political leaders were committed
to decentralizing planning and administrative functions, the
ability and willingness of the national bureaucracy to facilitate
and support decentralized development activities, and the
capacity of field officials of national agencies and departments
to coordinate their activities at the local level were strong
influences on decentralized development programs in nearly all
the cases examined.

The degree to which national political leaders supported
and focused attention on decentralized programs seems to
have had a profound influence on program implementation in a
number of cases. The Rural Employment Generation Program
in Thailand attained many of its goals because it had the
special attention of the Prime Minister, who chaired its

national committee, which included heads of the national ministries and departments whose support was needed to make the program operate effectively at the Province and Tambon levels. The strong interest of the Prime Minister made the program of high priority to cabinet ministers [52].

Central political support was also crucial in initiating district level coordination of agricultural programs in Sri Lanka in the late 1960s. District coordination was only successful as long as the Prime Minister gave it his personal attention and hand-picked senior administrators to serve as Government Agents. When he turned his attention to other matters, when senior administrators returned to the capital a year or two later and were not replaced by people of equal status, and when high level officials' 'monitoring visits' became less frequent, the ability of Government Agents to coordinate the activities of national departments within the districts waned quickly [53].

Some of the success of Pakistan's Integrated Rural Development Program in Manawala Markaz can also be attributed to the attention it received from high level political leaders and officials. Frequent visits by the national elites and the representatives of donor agencies created necessary compulsions for the national departments to demonstrate their commitment to the project by opening up their offices at or near the markaz complex. As the project manager got better access to his senior colleagues, logistical support to this markaz also improved. These factors helped Manawala Markaz to attract more institutions and physical infrastructure to its area and that way stay in the lead [54].

Another frequently cited factor in the ability of governments to implement decentralization was the willingness and capacity of national agencies and departments to support decentralized administrative units and to facilitate the coordination of development activities at the local level. Decentralization was underlined in Sri Lanka because the national civil service opposed arrangements that threatened its power and control. The civil service unions protected the prerogatives of central administrators and intervened actively in the political process to prevent a diffusion of administrative responsibility. Wanasinghe points out that:

> The general thrusts of these interventions have been towards maintaining individuality and autonomy of respective departmental cadres, strengthening the role of the bureaucracy in decision making, and enhancing career prospects through island-wide services. These thrusts have continuously run counter to attempts at implementation of local area-focused coordination, delegated decision making by peoples representatives, and creation of self-management organizations with their own personnel [55].

As a result, even when strong pressures came from the Prime Minister to coordinate activities within districts, many field officers resisted and 'the technical department cadres continued to maintain their allegiance to their own departments rather than to the district organization' [56]. Field officers considered the district a temporary assignment and their commitment was to national headquarters at which decisions about their promotion, salaries and assignments continued to be made.

The Rapti Integrated Rural Development Project in Nepal was also affected by the reluctance of national ministries to coordinate their activities within districts. Although the Chief District Officer was responsible for coordinating the operations of central line agencies, the control of central departments and ministries remained stronger, and each ended up acting separately and sometimes in isolation of the others [57].

Even in areas where Pakistan's IRDP was successful, the support and cooperation of the national departments remained weak. In Manawala and Harrappa representatives of some agencies such as the Irrigation and Industries Departments, the Agricultural Development Bank and the Punjab Agricultural Development and Supply Corporation rarely attended meetings of the markaz council. The Project Manager could get little support on technical matters from the national departments and often the kinds of advice they provided were inappropriate for the small scale development projects being undertaken at the local level. In neither markaz did representatives of most national departments receive adequate financial resources or transportation from their headquarters to be able to attend frequent coordinating meetings or to integrate their activities in the field. The IRDP approach did not fit well into the operating procedures of most national agencies and they made little effort to change their ways of doing things to facilitate or support district development planning and administration. Khan notes that they 'only associated themselves with this program as unwilling partners' [58] and that

> Each department and agency continues to pursue its program independently. Even the farm inputs, credit and agricultural extension agencies have not been able to properly integrate their field operations. Farmers from the two case study areas reported that they have to approach each of these agencies separately to avail themselves of their services. These farmers also mentioned that they do not look towards the Markaz as a source of supply of farm inputs and supporting services. They continue to go to the same old sources for meeting their needs to which they used to go before the introduction to the IRDP in their area [59].

ADMINISTRATIVE DECENTRALIZATION

Under these circumstances the success of IRDP depended almost entirely on the participation and cooperation of local leaders.

Behavioral, Attitudinal and Cultural Factors
Effectiveness in implementing decentralized programs depended in every country on behavioral, attitudinal and cultural factors, among the most important of which were the commitment of local officials to decentralizing development, the quality of local leadership, the attitudes of rural people toward government, and the degree to which traditional customs and behavior were compatible with decentralized administrative arrangements.

In a number of cases - SFDA in India, IRDP in Pakistan, the Provincial Councils in Sri Lanka and FAMA in Malaysia - the centrist attitudes and behavior of national government officials were revealed in their unwillingness to give local administrators discretion in carrying out local development functions. In the case of FAMA in Malaysia, State and district officers of the Authority were given virtually no autonomy in making decisions, even though they were dealing with unique and quickly changing conditions. Their lack of control, or even influence, over the prices they paid for crops or the disposition of the products they acquired, severely constrained their ability to react flexibly and effectively in carrying out the Authority's mandate. The Assistant State FAMA Officers, who worked at the district level in daily interaction with the farmers, could only make recommendations to FAMA headquarters in Kuala Lumpur, where all operating decisions were made. Thus, 'although he is, to all intents and purposes, the chief businessman for the [Agricultural Marketing] Center', Nor Ghani points out, 'he cannot conduct business according to his own terms and must continuously be guided by FAMA headquarters in Kuala Lumpur' [60].

The success of the Indonesian Provincial Development Program in Madura depended in large part on the willingness of provincial and local officials there to take on the additional work entailed in making the decentralized system work effectively. The PDP and especially the Small-Village Credit Program were very labor-intensive and the officials' time had to be allocated to a large number of new and unfamiliar tasks, including initiating 'bottom-up' planning, selecting target group participants, setting up new administrative arrangements that are responsive to local conditions and needs, training local leaders to manage the programs, implementing the projects, monitoring and evaluating their progress and preparing requests for reimbursements [61].

But in most countries, field officers or local administrators were reluctant to take the initiative in dealing with development problems, to exercise their leadership or to perform their tasks innovatively. This was due in part to the dependency of local officials on central government agencies and in part to social or cultural factors. Field officers of national ministries and local officials played a passive role in the Rapti Project, preferring foreign advisors to take the initiative:

> Except in a few cases, the line agency people are not found playing the role of leaders, rather their role has been played by the experts and advisors attached to the program. There is a general practice among villagers: they like to follow the instructions given to them by foreigners rather than those given by Napali citizens. It is their belief that the foreigners know more than the local people [62].

In India, those who managed the Small Farmers Development Agencies were given little discretion by the central bureaucracy - nearly all procedures and activities were prescribed by central rules and regulations - and there is little evidence that they were willing to innovate or take risks even when the opportunities arose. Mathur found that 'even where the agency is exhorted to adopt its own methods as determined by local conditions, it chooses to work in the well-trodden party of central guidelines'. In Alwar, for example, all of the projects were adopted 'from the shelf of schemes provided by the government' [63].

In Fiji, central officials distrusted the ability of Provincial Councils to make important decisions and often took unilateral action or by-passed them in carrying out rural development programs. For instance, the government's popular self-help rural development program was financed directly by the center with one-third of the funding coming from the villages. Ali and Gunasekera note that 'the government has deliberately left out the Provincial Councils both in disbursing its funds to the villages and in assigning priority to the projects to be implemented. The Central Government's view seems to have been that Provincial Councils are incapable of spending large amounts on capital projects' [64].

The behavior of local leaders was extremely important in nearly all of the cases examined. In the villages in Thailand that were successful in implementing the Rural Employment Generation Program, the relationships between the chief district officer and Tambon Council leaders were strong and the Tambon leaders played a vital role. Where local leaders were not skilled, honest and willing to commit their time and energy to initiating, following through on obtaining approval, and supervising the projects, they often failed or the financial allocations to the village were squandered [65].

The success of the Indonesian Provincial Development Program in Madura was also attributed to the positive attitudes of and strong leadership by provincial and local leaders. Where village leaders were willing to take initiative and to cooperate with higher level officials, the program usually succeeded. In areas where the small scale credit schemes worked best, 'the village head plays more than a pro forma role and does the groundwork for the operation of the program' [66]. Village leaders explain the objectives of the program and their rights and obligations to the villagers, and only after these preparations have been made do village leaders, the officials of the small village credit program, and other community leaders - with some guidance from the sub-district head - select those to receive loans. This open, participatory and cooperative process accounts for the smooth operation of the program and the good record of loan repayments in successful villages. 'The result is remarkable', Moeljarto observes. 'The commitment of all [local leaders] as manifested in their cooperative behavior and shared responsibility becomes one of the keys to the success of the KURK program' [67]. In villages where this leadership and co-operation were lacking, loans were often made on the basis of favoritism and to high-risk borrowers, resulting in many bad debts.

The success of the small village credit program in many communities in Madura was also due to the ability of officials to attain the cooperation of influential informal leaders and to overcome, ameliorate or avoid the potentially adverse effects of traditional behavior. In Madura:

> ... informal leaders are almost identical with kyais, i.e. religious leaders who, while not formally a part of the bureaucracy, play an influential role in society. They are very knowledgeable in Islamic dogma and religious rituals and constitute the desa [village] elite. The more orthodox kyais tend to perceive paying interest as haram or religious taboo. The provincial government is very well aware of this religious prohibition as interpreted by orthodox kyais, and therefore, prefers to use the term 'management fee' instead of interest. Even so, not all religious leaders buy the idea. Close and continuous communications and cooperation between KURK [Small Village Credit Program] officials and religious leaders is indispensable for the success of the program [68].

But decentralization in some countries also allowed local political leaders or elites to capture or dominate the programs for their own ends. In Pakistan, local political interests strongly influenced the process of resource allocation for IRDP projects. Some chairmen of Union Councils who were members of the Markaz Council attempted to get proposals for their own villages approved, but were apathetic to or opposed projects for other villages [69]. In Sri Lanka, district development planning was weakened during the 1970s because members of parliament and other influential politicians were able to exercise strong influence in selecting projects financed from the district budget. Technical factors and feasibility assessments were often disregarded. 'The involvement of appointed party cadres was perceived as bringing about participation of the people in development decision making', Wanasinghe noted. 'Narrow political interests favoring specific projects were perceived as articulation of public demand. Aggregation of individual electorate preferences in projects was perceived as district planning'. As a result the administrator came to be viewed as subservient to narrow political interests and 'recruitment to administration was perceived of as a logical compensation for loyalty to the party' [70].

Organizational Factors

Such organizational variables as the clarity, conciseness and simplicity of the structure and procedures created to do decentralized planning and administration, the ability of the implementing agency staff to interact with higher level authorities, and the degree to which components of decentralized programs were integrated, also influenced their outcome.

The difficulty in coordinating the activities of local administrators and national department representatives has already been described. Part of the problem was due to the low status of officials placed in charge of the programs. The staff of the Small Farmers Development Agency in India did not have fixed tenure and 'responded to the erratic government policy of postings and transfers' [71]. The program was headed by the District Collector and was only one of the many activities for which he was responsible. The government deliberately organized the program in a way that would keep it dependent on the central ministries, and 'the result was that partly by design and partly by environmental influence, the unique characteristics of the new agency never attained sharp focus' [72].

In Pakistan, IRDP Project Managers had virtually no formal powers to compel cooperation by representatives of national departments. Whatever success they achieved at the markaz level depended on their individual skills in persuading field officers and heads of local organizations to participate in Council activities and not on an organizational structure that facilitated or required integration of national efforts at the local level.

Where decentralized programs were organized in a way that made their purposes, structure and procedures clear, concise and uncomplicated, they seemed to have been much more successful than where the purposes were ambiguous and the procedures were complex. For example, the goals of the SFDA in India were overly ambitious and its procedures were difficult to apply at the local level. As the program was designed, about 50 million households would be eligible for aid. But at the rate at which SFDA was able to provide assistance Mathur estimates that it would take about 50 years to reach the beneficiaries; if population growth in the target group was considered it might take 150 years. Moreover, there was a large gap between central planners' rhetoric and what central officials were willing to allow local administrators to do. 'At the central level the planners usually talked in high ideal tones and insisted that the local level officials needed to respond to local situations and not to central instructions' Mathur contends. They told local officials that the most important goal of the program was to raise the income of beneficiaries and that loans and subsidies were only a means to that end and not ends in themselves. But 'these ideas somehow failed to percolate down' he argues [73]. In reality, SFDA staff were shackled by detailed central rules and regulations, many of which were inapplicable at the local level. Evaluations were based on the number of loans made rather than their impact on beneficiaries. The banks that made the loans to small scale farmers were more concerned with repayment than with the effects on agricultural production.

The difficulties of implementing decentralization in Sri Lanka can also be attributed to ambiguity in design and organization. Purposes of the district budget were never clarified and as a result varying interpretations emerged, 'ranging from that of the provision of an electoral fund to the dawn of a district planning and budgeting exercise'. Wanasinghe points out that 'this resulted in confusion, with the more conscientious elements in the bureaucracy attempting to inject techno-rationality into the program and the Member of Parliament seeking to entrench his or her position in the electorate through distribution of favors' [74]. Moreover, the district budget process was never well integrated with the coordination functions of the Government Agent and as a result 'the whole issue of providing resources to match decentralized responsibilities remains unresolved' [75].

The ambiguities remained in the design of District Development Councils established in 1980. The relationships between the Councils and national agencies, for example, were left undefined. The District Ministers appointed in 1981 were not from electorates in the districts to which they were assigned. Wanasinghe notes that 'they derive their authority from the Executive President, owe their tenure to him, are not recallable by the people of the districts they are ministers of, and, on these counts if they are agents of anyone they are agents of the government at the center rather than the people at the periphery' [76].

In Indonesia and Thailand, however, clearly defined purposes and procedures allowed programs to progress more smoothly and effectively in many areas. Chakrit points out that 'the central government has carefully laid down the responsibility and the authority, as well as the expected roles of respective levels of government' [77]. In Indonesia, rules and procedures were realistic and applicable at the local level. Province officials guided and supervised the program to ensure that it was carried out effectively, but left room for local initiative and flexibility. Moeljarto comments that the 'frequency of visits of provincial BAPPEDA and sectoral staff from Surabaya to Madura for guiding, supervising and monitoring activities seems to be high' and that this not only motivated local officials but created a system of checks and balances that maintained effective implementation [78].

Financial, Human and Physical Resources
The UNCRD cases also underline the crucial importance of adequate financial resources, skilled personnel and physical infrastructure at the local level.

The adequacy of financial resources and the ability to allocate and expand them effectively were noted in nearly every case. The lack of independent sources of revenue weakened the SFDA's ability to carry out its tasks in India. The dependence on central government grants kept the SFDA under the control of the central bureaucracy. Even in countries such as Fiji that devolved revenue raising powers to local governments, localities remained dependent on central funding for the bulk of their activities. Ali and Gunasekera point out that after more than a decade of devolution in Fiji, the Provincial Councils still receive about 55 per cent of their revenues from central government grants [79].

In some countries it was not the lack of financial resources that created problems but the inability of decentralized units to spend the money they received. In Nepal, nearly 80 per cent of the funding for the IRDP in Rapti came from USAID and there was no dearth of financial resources. The problem was getting the USAID funds, which were channeled through the Ministry of Finance in Kathmandu, to the Project Coordination Office in a timely manner. In fiscal year 1981-2, for example, it took five months for allocated funds to reach the project. The central ministry imposed its standard rules on expenditure transactions and did not allow project managers to transfer funds from one budget to another. When funds were late reaching the PCO, projects were delayed or postponed until money became available.

In Fiji, cultural factors inhibited Provincial Councils from using their taxing powers. The difficulty they had in collecting taxes was not due to the lack of resources within the provinces but to the strong communal structure that encouraged individuals to pay taxes only for activities that benefitted the entire community. 'It seems that villagers are willing to incur costs such as payment of rates only if such funds are used for visible projects leading to the welfare of all the people within the Provincial Council area, rather than one group of people', Ali and Gunasekera observed [80]. In any case, Provincial Councils had no practical way of enforcing tax laws. Communal living did not allow private property to be identified easily, and even if property was seized to pay the taxes it would be difficult to resell. Moreover, the pressures on elected Councillors against seizing property to pay taxes would inhibit them from doing so. Thus, most provincial councils did not even keep records of tax payments.

But because the communal tradition also obliged people to assist in projects for the common welfare, many were willing to make voluntary contributions to the Provincial Councils for such activities. In some areas the Councils were able to make up the short-falls in tax collections with voluntary contributions.

An equally important factor influencing program implementation was the availability of skilled staff at the local level. Many programs were plagued with shortages of trained technicians and managers. The SFDA in India was especially weakened by the rapid turnover of personnel within districts. In Alwar, the average tenure of the District Collector - who headed SFDA - had been 17 months, and of the project officer 18 months. Although one project officer stayed for 48 months, this was rare, and during his term there were three changes in the District Collector, two in the agricultural project officer, two in the animal husbandry officer and five in the cooperatives officer. The knowledge that posts were temporary gave local officers little incentive to take responsibility for their functions or to build effective teams to coordinate their activities [81].

Similarly, in Sri Lanka the officers assigned to the districts saw them as temporary appointments which they would hold only until they could get an assignment in the national capital, and were not willing to take risks or make mistakes that would threaten their promotion or reassignment.

In Pakistan, the technical personnel available to the Markaz Councils were quite limited. 'Only one sub-engineer is attached to the Project Manager' Khan observed. 'It is rather difficult for one person to look after the development work in all the 50 to 60 villages falling in the Markaz territory' [82]. Moreover, the Project Managers were often inadequately trained to do their jobs. Most were either agricultural technicians or generalist administrators who had little or no experience with area-wide planning and development.

The Rapti project in Nepal was also constrained by the lack of trained technicians and managers and by the unwillingness of those with adequate training to live and work in the region. Ninety per cent of the project office staff were low-level, ungazetted, employees, who were capable of carrying out only the most basic and routine tasks [83].

Finally, physical conditions and the adequacy of physical infrastructure seemed to affect the ability of field officials to implement decentralized programs. Many of the areas in which the program was set up were rural regions or provinces remote from the national capital and in which settlements were not linked to administrative centers. The cooperation and interaction envisioned in the design of decentralized programs were difficult to achieve because of poor roads, lack of transportation to villages and towns, and poor communications systems. National department representatives had found it difficult to travel to the markaz for meetings or to the 50 to 60 villages that were within the jurisdiction of the markaz centers. The Rapti zone in Nepal was physically isolated from the national capital; transportation facilities between the headquarters of the Project Coordinating Office and districts in the Rapti region were extremely weak and communication was one of the biggest problems in expediting the program in the zone [84]. The Project Coordination Office, located in Tulsipur, was 20 kilometers from Dang and the other four hill districts were farther away. Nearly all of the communications took place through official correspondence or, in urgent situations, by wireless because of the difficulties of travelling from one part of the region to another or between the region and Kathmandu. During the monsoon season it could take up to four days to travel from one district to another and in this period there was virtually no communication between the PCO and the line agencies.

CONCLUSION

Thus, although decentralized programs were initiated in Asia and the Pacific for quite different reasons and have taken different forms in different countries, many of the problems that arose in formulating and implementing them seem to be common and recurring [85]. It is to these problems - increasing political and administrative support for decentralization from the center, organizing programs in ways that are conducive to field management, creating or changing attitudes and behavior of central officials, field staff and rural residents toward decentralized planning and management, and providing adequate financial, human and physical resources at the local level - that the attention of international assistance organizations and governments of developing countries must turn if such programs are to be carried out more efficiently and effectively in the future [86].

Clearly, the administrative capacity of local organizations must be strengthened before new functions and responsibilities are assigned to them [87]. The tasks of central ministries and agencies must be reoriented in a decentralized system of administration from control to supervision and support, and their capacity to strengthen local government or administrative units must be expanded. Decentralization holds new opportunities and responsibilities for both local administrators and central bureaucracies, but it will not succeed unless they mutually support and reinforce each other. Finding ways of building the capacity of local administrative units to implement development programs and of eliciting the support of central bureaucracies in that task offers an important challenge to governments of developing countries in the years to come.

NOTES

1. For a comparison of these results in Asia with those in other developing regions see G. Shabbir Cheema and Dennis A. Rondinelli (eds) Decentralization and Development: Policy Implementation in Developing Countries (Beverly Hills: Sage Publications, 1983), in press.

2. The papers were commissioned as part of the Project on Implementing Decentralization Policies and Programs.

3. See Dennis A. Rondinelli, 'National Investment Planning and Equity Policy in Developing Countries: The Challenge of Decentralized Administration', Policy Sciences, Vol. 10, No. 1 (1978), pp. 45-74.

4. For a more extensive discussion see Dennis A. Rondinelli, 'Administration of Integrated Rural Development: The Politics of Agrarian Reform in Developing Countries', World Politics, Vol. XXXI, No. 3 (April 1979), pp. 389-416.

5. Kuldeep Mathur, 'Small Farmers Development Agency in India: An Experiment in Controlled Decentralization', paper prepared for Senior Level Seminar on Implementing Decentralization Policies and Programs (Nagoya, Japan: United Nations Center for Regional Development, 1982), p. 10.

6. Ibid., p. 15.

7. Dilawar Ali Khan, 'Implementing Decentralization Policies and Programs: A Case Study of the Integrated Rural Development Program in Punjab, Pakistan', paper prepared for Senior Level Seminar on Implementing Decentralization Policies and Programs (Nagoya: UNCRD, 1982), p. 51.

8. Ibid., p. 3.

9. Ibid., p. 2.

10. Moeljarto Tjokrowinoto, 'Small Village Credit Program of the Provincial Development Program in East Java', paper prepared for Senior Level Seminar on Implementing Decentralization Policies and Programs (Nagoya: UNCRD, 1982), p. 31.

11. Chakrit Noranitipandungkarn, 'Creating Local Capability for Development Through Decentralization Programs in

Thailand', paper prepared for Senior Level Seminar on Implementing Decentralization Policies and Programs (Nagoya: UNCRD, 1982).

12. Shelton Wanasinghe, 'Implementing Decentralization Policies and Programs: The Sri Lankan Experience', paper prepared for Senior Level Seminar on Implementing Decentralization Policies and Programs (Nagoya: UNCRD, 1982), p. 6.

13. Ibid., pp. 17-22.

14. Government of Sri Lanka, Sessional Paper V of 1982, p. 32; quoted in Wanasinghe, op. cit., p. 38.

15. Ahmed Ali and H. M. Gunasekera, 'Implementing Decentralization Policies and Programs: The Case of Fiji', paper prepared for Senior Level Seminar on Implementing Decentralization Policies and Programs (Nagoya: UNCRD, 1982), p. 11.

16. Ibid., p. 18.

17. Mohd Nor Abdul Ghani, 'Management Decentralization: A Case Study of the Federal Agricultural Marketing Authority (FAMA) in Malaysia', paper prepared for Senior Level Seminar on Implementing Decentralization Policies and Programs (Nagoya: UNCRD, 1982), p. 2.

18. Gabriel U. Iglesias, 'PDAP: A Case Study on a Strategy for Strengthening Local Government Capability', paper prepared for Senior Level Seminar on Implementing Decentralization Policies and Programs (Nagoya: UNCRD, 1982).

19. Bhim Dev Bhatta, 'Rapti Integrated Rural Development Program', paper prepared for Senior Level Seminar on Implementing Decentralization Policies and Programs (Nagoya: UNCRD, 1982), p. 6.

20. For an amplification on this definition see Dennis A. Rondinelli, 'Government Decentralization in Comparative Perspective: Theory and Practice in Developing Countries', International Review of Administrative Sciences, Vol. XLVII, No. 2 (1981), pp. 133-45.

21. Mathur, op. cit., p. 15.

22. Chakrit Noranitipadungkarn, op. cit., pp. 7-11.

23. Khan, op. cit., pp. 26-7.

24. Ibid., p. 30.

25. Ibid., pp. 28-9.

26. Ibid., p. 35.

27. Wanasinghe, op. cit., p. 14.

28. Ibid., pp. 17-18.

29. Ibid., pp. 24-30.

30. Iglesias, op. cit.

31. Moeljarto, op. cit., pp. 3.7-3.10.

32. Ibid., p. 3.11.

33. Bhatta, op. cit., p. 8.

34. Ibid., pp. 15-20.

35. Mathur, op. cit., pp. 12-15.

36. Ibid., p. 42.
37. Nor Ghani, op. cit., pp. 7-8.
38. Wanasinghe, op. cit., pp. 40-2.
39. Ali and Gunasedera, op. cit., p. 22.
40. Mathur, op. cit., pp. 28-9.
41. Chakrit Noranitipadungkarn, op. cit., p. 29.
42. Ibid., p. 53.
43. Khan, op. cit., p. 65.
44. Ibid., p. 53.
45. Bhatta, op. cit., pp. 37-40.
46. Iglesias, op. cit.
47. Wanasinghe, op. cit., p. 16.
48. Ibid., pp. 23-7.
49. Ibid., p. 31.
50. Ali and Gunasekera, op. cit., p. 46.
51. Ibid., p. 47.
52. Chakrit Noranitipadungkarn, op. cit., p. 5.
53. Wanasinghe, op. cit., p. 21.
54. Khan, op. cit., p. 67.
55. Wansinghe, op. cit., pp. 49-50.
56. Ibid., p. 22.
57. Bhatta, op. cit., p. 55.
58. Khan, op. cit., p. 53.
59. Ibid., p. 73.
60. Nor Ghani, op. cit., p. 16.
61. Moeljarto, op. cit., p. 6.11.
62. Bhatta, op. cit., p. 47.
63. Mathur, op. cit., p. 68.
64. Ali and Gunasekera, op. cit., p. 38.
65. Chakrit Noranitipadungkarn, op. cit., pp. 60-1.
66. Moeljarto, op. cit., p. 5.18.
67. Idem.
68. Ibid., p. 5.20.
69. Khan, op. cit., p. 68.
70. Wanasinghe, op. cit., pp. 28-9.
71. Mathur, op. cit., p. 67.
72. Idem.
73. Ibid., p. 69.
74. Wanasinghe, op. cit., p. 44.
75. Ibid., p. 36.
76. Ibid., p. 39.
77. Chakrit Noranitipadungkarn, op. cit., p. 60.
78. Moeljarto, op. cit., p. 6.4.
79. Ali and Gunasekera, op. cit., p. 28.
80. Ibid., p. 32.
81. Mathur, op. cit., pp. 47-8.
82. Khan, op. cit., p. 50.
83. Bhatta, op. cit., p. 21.
84. Ibid., p. 46.
85. For a more detailed discussion see Dennis A. Rondinelli, John R. Nellis, and G. Shabbir Cheema, De-

centralization in Developing Countries: A Review of Recent Experience, World Bank Staff Working Paper No. 581 (Washington: World Bank, 1983).

86. See G. Shabbir Cheema and Dennis A. Rondinelli (eds), Decentralization and Development (Beverly Hills: Sage Publications, 1983).

87. This argument is made more extensively in Dennis A. Rondinelli, 'The Dilemma of Development Administration: Uncertainty and Complexity in Control-Oriented Bureaucracies', World Politics, Vol. 35, No. 1 (1982), pp. 43-72.

88. Dennis A. Rondinelli, Development Projects as Policy Experiments: An Adaptive Approach to Development Administration (New York: Methuen, 1983).

Chapter Three

THE POLICY ENVIRONMENT NECESSARY TO
MAKE EXTENSION EFFECTIVE

G. Edward Schuh
The World Bank

There is a little book called The Principles of Agriculture by
L.H. Bailey which is somewhat of a classic [1]. At the beginning of this book the author states, 'Again, the purpose of
education is often misunderstood by both teachers and
farmers. Its purpose is to improve the farmer, not the farm.'
There is a world of wisdom in that statement. It is striking to
note the extent to which the Land-Grant Universities have
strayed from that dictum and have worried more about crops,
animals and soils than about human beings or agriculturalists.
It is with this thought that one should approach a topic as
important as the policy environment necessary to make extension effective.

Extension programs are often developed as if economic
policy does not matter, and often as if economics does not
matter. However, beyond that point, one must address, as
well, the issue of extension's economic education program for
farmers. This tends to be a much neglected part of our
extension programs, yet it is every bit as important as the
technology side of our programs.

This paper is divided into three parts. The first is a
discussion of the policy environment necessary for extension
to be effective. The second considers the changed international environment for agriculture and some implications of
that changed economic environment. That provides a basis for
the third part of the paper, which has to do with policy
education per se. The discussion does not go into the
technical details, but instead focuses on basic principles.

---

*The views and opinions expressed in this chapter do not
necessarily reflect the position or policy of The World Bank
and no official endorsement should be inferred.

THE POLICY ENVIRONMENT FOR EFFECTIVE EXTENSION

The place to begin is to note that for economic policy to create a favorable environment for extension it has to make the adoption of new production technology profitable. It is well known that it is not automatic that new production technology will be profitable. Biological and physical scientists sometimes appear to believe that any new production technology will increase the farmer's income. However, since most new production technology is embedded in an input in one form or another, it follows that the price relative (the ratio between the product price and the input price) has to be such as to make it profitable to use the new or improved input. And if price relatives are not such as to make it profitable to use the new input, no amount of preaching or belaboring the farmer will have much of an effect.

Consider an example from Brazil. An important issue some ten years ago was why farmers were not using fertilizer on maize, but were using it on other crops. The extension service at that time believed that the use of fertilizer was rational for all crops for which there was a physical response. Extension workers recommended the use of fertilizer on maize, but were then disappointed because the farmer did not use it. If the farmers were queried as to whether they had ever used fertilizer on maize, they would almost always say 'Yes - the extension people said we should use it'. If they were then asked why they were not using it now, their answer was almost inevitably 'Nao compensa' (It doesn't pay!). As a consequence of these misguided recommendations extension was discredited. To the producer it appeared that the extensionists did not know what they were talking about.

Thus, the question of profitability is important. In an exaggeratedly simple way, this profitability can be obtained by having higher prices for the product and lower prices for the input. If there is a wide ratio between these two prices, an extension program can be very effective in promoting the use of fertilizers or other modern inputs.

There are a number of examples around the world where policy makers have attempted to obtain a wide gap between the product and input prices specifically for the purpose of promoting the adoption of a new technology. It is fairly easy in principle to obtain such a wider price ratio. The practical issue, however, is how far such changes in prices should go. What is the criterion for setting those prices so that it is profitable to adopt the new technology?

When considering products and inputs that are traded on international markets, there is again a fairly simple answer (at least in principle). The goal of price policy should be to make the most efficient use of the nation's resources. It is true that price policy can be used to change the distribution of income. But in general, price policy is not an effective means of accomplishing that. This is not to argue against changing the distribution of income – if societies want to change the distribution of income, that is their choice. The point is that there are more cost-effective ways of doing it than by changing price relatives. Hence, in establishing product and input prices it is better to focus on efficiency issues alone.

Using a nation's resources most efficiently can be accomplished by setting domestic prices equivalent to border price-levels. Hence, when considering something that is an import, which is often the case with fertilizer, one would put the price at a level consistent with the CIF import price at the port of entry. When considering an export, which is often the case on the product side, then the FOB price at the port of export is taken as the base. In each case, of course, it is necessary to adjust for transportation costs within the country.

When one suggests this criteria for establishing prices many people argue that domestic prices for something as important as food should not be dictated by international market conditions. There are exceptions to the rules, of course, but exceptions to the rules should be permitted only under conditions in which it really does make sense.

A frequent argument is that farmers have to be sub-sidized to promote the adoption of new technology. The logic of the argument is that the adoption of technology has a high payoff to society and that a farmer should be compensated for the risk he or she incurs in using the new or improved input. Under certain circumstances, this argument has some validity. Technically it is correct. But again, one should be careful not to overdo these kinds of arguments, in part because it is difficult to remove subsidies once they are in place.

The best way to gain perspective on this proposition is to cast the problem in a larger context. By so doing, we immediately recognize certain adjustment costs that are associated with the adoption of new production technology. The bulk of these costs are generally borne by the agricul-tural labor force, typically those with a lower income. If one accepts the criteria laid out above in very simple terms, then the final outcome is an optimal rate of technical change that will most likely minimize adjustment costs. This outcome will generally be in marked contrast to that which is obtained when the adoption of new production technology is sub-sidized.

Having noted this, it is fair to add that in most devel-oping countries the pursuit of policies which induce an excessive rate of technical change is not likely to occur. To the contrary, in most developing countries economic policy severely discriminates against the agricultural sector and against the adoption of new technology. It discriminates against the agricultural sector by doing just the opposite of what has been discussed; product prices are pushed below their international border price-levels, while at the same time the prices of modern inputs are fixed above their border price equivalent. The latter is usually done as a means to promote the development of the input supply industry. Unfortunately, this again is a policy that sounds good in principle, but which once in place is very difficult to remove.

Another related point is that policies which discriminate against agriculture also lower the rate of return on investment in research and extension. Plausible as that may seem, one of the frustrating things about policy-making is the general failure to recognize the extent to which there are spill-over effects from distortionary policies. When price relatives are distorted against agriculture, for example, it is not only that the output from the given set of resources is reduced and that land, labor, and other inputs used in the sector are thus under-valued. The rate of return to investments in research and extension and in educational programs is also reduced and the investments themselves are under-valued. Unfortunately, it is not only the policy-makers who fail to recognize this problem. Professional observers also frequently overlook it, and consequently criticize extension and research programs for not being effective. They have failed to note that the price and policy environments discriminate very severely against those investments.

There are a number of other issues that need to be addressed in this context. The first is the issue of credit policy. Again, few things tend to be more distorted in developing countries than agricultural credit policies. Rather than let markets evolve and develop, countries tend to impose barriers and to intervene in various ways that impede the mobilization of savings for reinvestment in agriculture, and distort the use of credit within the agricultural sector (or in agriculture vis-à-vis the rest of the economy). These interventions range from usury laws, to sectoral allocations of credit by central planners, to administrative allocations of credit within the agricultural sector in favor of food crops and against export or cash crops, to distortions in the terms on which credit is provided.

One of the common interventions in the case of agriculture proper is the tendency to have cheap credit to subsidize the use of modern inputs. Worldwide, that may well be the most common distortion affecting the economic environment for extension programs - with very significant impact. This kind of intervention typically has two justifications. The first is the use of cheap credit to offset the effect of discriminatory policies that are reflected in price relatives. That is a very common justification. The second justification is the avowed need to stimulate the adoption of new technology. It is typically believed that cheap credit is needed to induce farmers to adopt a new technology, and that once they are 'hooked', they will continue to use it.

Two comments are relevant on such justifications for cheap credit. The first is a reminder that when one sub- sidizes credit there almost always has to be some means of rationing the credit. When policy-makers turn to non-price kinds of rationing, personal influence tends to determine who receives credit. That means that the recipient is not likely to be the poor small producer who is suffering from capital rationing. The people who receive the credit under these circumstances tend to be the large producers - those who would generally receive the credit under any circumstances. It is not clear that large producers will make most effective use of the credit and in turn facilitate the adoption of the production technology.

The second point is that the effects of such subsidies tend to get capitalized into the value of the land. That has very important income distribution consequences which may not be in the original intended direction of policy. One of the things to note, of course, is that the capitalization into land values can have an effect on the adoption of technology in its own right, since it makes the price of land more expensive relative to land-substitutes like fertilizers, pesticides and herbicides. This is a second order of things compared to what one would originally expect from the use of credit.

In conclusion, the burden of this discussion is to argue that we should think twice before we select credit subsidies as the means to make extension more effective. Instead, we should focus on getting the basic price ratios where they should be in the first place.

Another aspect of the economic environment which does not receive adequate attention is the policy environment necessary to deal with environmental problems and sustain- ability. The environmental degradation that occurs in much of Africa and in other parts of the world is especially trouble- some. Much of this is induced by bad policy, and is not something that would require an explicit subsidy to eliminate.

Two aspects of economic policy contribute to environ- mental problems and the lack of sustainability of a modern agriculture. One is policy which discriminates against agri- culture and thus causes the resources in agriculture to be under-valued. If land and natural resources are under- valued, the incentives to husband these resources are obviously less than if the proper price relative were to prevail. The second issue is the failure to properly price certain inputs - the tendency to make them free goods, which implies an infinite demand for them - and the corresponding lack of incentives for efficient investment in such inputs. Two examples come to mind: (1) water, which tends around the world to be zero priced to the farmer; and (2) the tendency in much of Africa and other parts of the world to make firewood free to people and then to lament the fact that they slash it, burn it, cut it down, and thus destroy whatever soil-holding power the forests might have for agriculture itself.

These issues are important, and seldom are addressed because we tend to focus on making the crop grow better and the cow look prettier. We need to give greater attention to such things as the pricing of water. It is not an easy thing to do in many cases, but there are ways of approximating market prices. When water is properly priced, more efficient use is made not only of the water, but of all the other inputs used with it. The same situation prevails with the problem of firewood in Africa. When firewood is free, there is no incentive to invest in alternative sources of energy. Again, as in the case of water – the main consideration is not what is done about the direct use of the firewood, but the implications of the policy for investment policy in firewood plantations and other related activities.

THE CHANGED INTERNATIONAL ENVIRONMENT FOR AGRICULTURE

Examination of agriculture in an international context raises another important set of issues. Agriculture is the most well-integrated sector of the economy internationally, and not only in the case of the United States. Almost all countries either export or import some agricultural product, and in many cases they do both. For example, the United States is a large exporter, but it is also the second largest importer of agricultural products in the world. There is a great deal of difference between thinking about agriculture as open to the international economy and thinking about it as a closed economy. Thinking about agriculture as a closed economy when it is in fact open is a trap, and leads not only to some of the bad policies we now have globally for agriculture [2] but also to the failure to recognize the proper criteria for establishing agricultural prices.

International trade for agricultural products grew very significantly in the 1970s. It has since declined, but the significance of international trade prevails despite the fact that agriculture probably has more distortions to trade than any other sector of the economy, and despite the fact that agriculture has not benefited from the trade liberalization associated with the multi-lateral trade negotiations of the post-World War II era.

Secondly, it is important to emphasize the significance of the well-integrated international capital market which now links most countries of the world in ways as important or more important than international trade. If one goes back to the end of World War II, there was virtually no international capital market. There were some transfers of capital among countries, but it was mostly done on a government to government basis, and we called it foreign aid. If one goes much before World War II, there were rather large flows of international capital, but nothing like what we have today. As we moved through the 1960s and 1970s, we successively have witnessed a Euro-dollar market, a Euro-currency market, and then a flood of petrodollars. By 1984 (the most recent year for which data are available), we find that total international flows of capital had risen to 42 trillion dollars, while international trade was only two trillion dollars. Thus we see that the trade side of our economic relations is minuscule compared to the international capital market. Indeed, what is causing exchange rates to fluctuate so much has very little to do with trade. Those who have studied international trade probably believe that exchange rates are determined by export and import performance, but the real driving force today is what is happening in the capital market.

It is difficult, in fact, to overstate the significance of the international capital market. We have become very cognizant of it as reflected in the international debt crisis. But again, that is a fairly small part of the picture. When such an international capital market is combined with the kind of floating exchange rate system we now have, there emerges a very strong link between financial markets and commodity markets. For example, it is not necessary to consult the commodity page of the Wall Street Journal to know what is happening in commodity markets - only to read the financial page. There is also a very strong link between monetary and fiscal policy and commodity markets - something we have not had before. Together with these linkages, we also have strong linkages across countries. U.S. monetary policy also has a great influence not only on U.S. commodity markets, but on markets in other countries as well. That side of the story has simply not been told very often in recent years.

The point of this part of the discussion is to underline the need to consider the linkages between monetary conditions and commodity markets. Very large swings in exchange rates tend to obliterate any underlying comparative advantage. This masking of comparative advantage raises a number of difficult questions about where to invest in creating new production technology, and about the proper economic policy for an individual country.

ECONOMIC POLICY EDUCATION

Finally, it is important to emphasize the need for a great deal more policy education in extension. We need to move extension programs away from the production technology side alone and do a better job on policy education. First, we should stress the importance of policy in determining the welfare of rural people, in determining the profitability of their agriculture and, in turn, in determining the adoption of specific production technologies. Second, we should consider changes in the configuration of the international economy which (in light of the fact that agriculture is a truly globally integrated sector of the economy) has enormous implications in almost every country. Third, policy education is a very important part of developing a democratic society in which our citizens are able to choose the kind of policies that serve them best. Fourth, producers cannot really make economically intelligent decisions about their own operations without understanding what these policies are.

CONCLUDING COMMENTS

In conclusion, there is an obvious complementarity between science and technology policy, on the one hand, and economic policy on the other. Without a proper economic environment, research and extension programs will simply not be very effective. They will have very low social rates of return, except under unusual conditions. The incentives have to be present in order for producers to adopt the new technology being extended.

There are two corollaries that follow from the previous statement. The first is the need for more effective policy education programs. The second is the need to cast that policy education in the context of the international economy, of which agriculture is a part. In today's world we can hardly do anything else if we want to be relevant.

NOTES

1. L.H. Bailey, (1918) The Principles of Agriculture. McMillan Company, New York.
2. For an overview of how distorted agricultural policies have become around the world, see World Development Report. The World Bank, Washington, DC (1986).

Chapter Four

AN OVERVIEW OF AGRICULTURAL EXTENSION AND ITS
LINKAGES WITH AGRICULTURAL RESEARCH:
THE WORLD BANK EXPERIENCE*

Donald C. Pickering
The World Bank

INTRODUCTION

This discussion provides an overview of agricultural extension
and its linkages with research from the perspective of the
World Bank, with emphasis on the African situation. Although
the text borrows heavily from material prepared and
presented at the series of Workshops on Extension and
Research sponsored by the World Bank and other agencies
over the past four years, the same observations are as true
today as they were in 1982, 1983 and 1984. Their force can
only be enhanced by reiteration.

There has probably never been a time when promoting
increases in the productivity of sub-Saharan African agriculture was more important or more clearly comprehended. The
continent is characterized by rapidly rising populations; local
production of basic foodstuffs cannot keep pace with population increases in many countries; agricultural exports have
declined overall, with some exceptions; the area has suffered
and is still suffering from widespread adverse weather conditions; and reserves of unused potentially productive agricultural land have been severely depleted in many countries.

Sub-Saharan Africa depends more than most other
regions upon its agriculture for economic growth and the
well-being of its people. It is obvious that if such growth is
to be attained and if its people are to achieve higher living
standards, agriculture must become more productive. This is
more easily said than done.

An historically informed review of West African agriculture indicates that many different systems of agricultural
research and extension have been preached, and some have

---

*The views and opinions expressed in this chapter do not
necessarily reflect the position or policy of the World Bank
and no official endorsement should be inferred.

66

been practised in the region over the years. They were designed with different and sometimes unclear objectives in mind, in different agro-ecological zones, with different crop or livestock enterprises and applied in (but not necessarily designed for) varying socio-economic circumstances. This said, there is no single blueprint for 'the best' research or extension approach, since any and all have to take into account the context and conditions under which they must operate. While there is no blueprint, there are some basic principles that need to be respected, well defined, and adequately translated into operational principles.

Over the last eight or ten years the World Bank has given a great deal of attention to agricultural extension and research as major avenues for channelling technical assistance to large numbers of small farmers. Within the Bank's overall lending for agriculture, and particularly for rural development and poverty alleviation, substantially increased financial and staff resources have been directed to improving extension services and strengthening agricultural research. The Bank has continued to give strong support and impetus to the search for innovative approaches in extension, to the efforts for eliminating the chronic organizational weaknesses that have plagued extension services, and to sound policies for linking agricultural research to extension work and to farmer needs. Support has been given for various and differing approaches to extension, adjusting programs to suit local conditions and resources.

The Bank has gone through a continuous learning process, based on critical assessments of actual experience and on avoidance of blind acceptance of one or another 'blueprinted' system. At this point it may be helpful to examine some of the different approaches to extension in Africa over the recent past, explore their relative strengths and weaknesses, and derive appropriate lessons for current and future activities.

EXTENSION APPROACHES

Without attempting to provide definitive classification or categorization, this discussion reviews several different extension approaches that have been attempted. These approaches have been adapted from Six Approaches to Rural Extension, a paper by B. Haverkort and N. Roling, that is used at the International Agricultural Center in Wageningen for its International Course on Rural Extension. They have evolved historically and have changed over time in one or another respect, so that they are rarely found now in practice in a 'pure' form. Also, some of the approaches are partially overlapping and certain elements have been transferred from one to another. They include:

- the commodity-focused approach in extension;
- the community development-cum-extension approach;
- the technical innovation centered approach;
- the training and visit system approach;
- the 'animation rurale' approach;
- and several, more or less overlapping, other approaches.

One of the most widely spread formal extension patterns in West Africa has been the commodity-focused approach, shaped as a set of procedures designed to facilitate the production of a single crop (usually one not used for subsistence). The approach is based on the technical, administrative and commercial requirements of this crop, and is managed by a parastatal board or society, or by a private company. Successful examples from Africa include notably the activities of the French based company, CFDT, with cotton in a number of francophone West African countries. The British American Tobacco Company provides another example in both East and West Africa.

This approach normally features a good technical package for the crop in question, systematically conveyed to farmers. The best examples are characterized by integration of extension advice with reliable input supply and with output marketing arrangements, together with prompt payment to farmers for their production. Properly operated such schemes can be very efficient because of the emphasis on cost effectiveness by their managements.

However, the commodity-focused approach often implies monopoly powers for the parastatal or crop processing and marketing organization, and carries the risk of facilitating excessive profits at farmers' expense. If poorly managed, or if changes in terms of trade and pricing affect the comparative advantages of the crop, poor returns to farmers can result. The emphasis on one crop can result in disregard of local needs, especially for traditional food production in the whole farm context. The best examples, however, increasingly take these factors into account, i.e. the cotton parastatals in Mali, Ivory Coast and Togo which have extended their crop coverage and now provide technical advice and credit for the key food crops grown along with cotton in the farming systems concerned.

The community development-cum-extension approach has operated to a limited extent in Africa and to a greater extent in other parts of the world, such as India. This approach is constructed around a rather broad definition of the functions of the extension agent. While positively attempting to link extension to other aspects of overall community development, this approach has diluted the specific agricultural extension responsibility of the village agent with a long, diffusely defined list of tasks. It has dispersed both his attention and his accountability among many different and non-focused activities. Although often called an 'extension agent', such a worker tends to be either a general community or general agricultural officer, simultaneously charged with administrative duties, family planning or health service, credit schemes, technology promotion, distribution of supplies, political mobilization functions, or ad hoc assignments such as census-taking, etc. This wide-ranging set of duties usually results in low performance, confused supervision, discontinuity, lack of mobility, little organized work, and an ineffective agricultural extension service. In a time when specialization and professionalization are clear prerequisites for technical progress in agriculture, such extension cannot significantly increase production and productivity levels.

The innovation-centered approach in extension regards its function primarily in terms of technology transfer from 'outside' to the farm, sometimes specifically in terms of 'selling' a number of technical innovations. The inherent problem that undermines this approach is its insufficient appreciation of the farmers' circumstances. Rather than starting from the farmers' conditions, and daily faced constraints, it starts from ready-made and outside packaged innovations, to be grafted into the socio-economic context of a farm which may not be capable of absorbing them. Another frequently encountered problem with this approach in Africa is the weakness of the technical information being extended, usually deriving from a failure to carry out the final stages of testing on farmers' fields in differing ecological zones and with different types of farmers.

The farmer-focused approach, as exemplified by the Training and Visit System, is an organizational approach which puts the farmer and his constraints, abilities and needs at the center of the whole extension effort. This approach mobilizes the entire extension apparatus and the research system to service the ultimate (small scale) producer. Properly operated, this approach also disseminates innovations and technical recommendations, but takes as its starting point the farm and its immediate difficulties and potential. It addresses both food and cash crops to the extent that relevant information is available. As a management system it tries to overcome the problems of the traditional governmental extension approach by promoting regularity of visits, regularity of training, effective supervision, and specialization of agents, focusing extension efforts on well researched, key impact points, and selecting contact farmers representative of different socio-economic groups of the farm population. Most important is sustaining and improving research/extension linkages through careful diagnosis of initial impact points, farm systems research investigations and joint conduct of adaptive trials on farmers' fields. Typically the system has been grafted to the innovation-centered approach, and there is a danger that it remains 'top down'. A conscious effort therefore has to be made to ensure that it is, in fact, farmer-focused.

A useful operational distinction can be made with regard to the farmer-focused approach to extension, between addressing the farmer individually (through the selection of contact farmers) and approaching groups of farmers (either already in existence or formed for extension purposes). There are significant methodological and sociological implications involved in utilizing one or other of these modalities of agent-farmer communication, and it is beneficial to discuss their comparative advantages or disadvantages and their relative adequacy in different cultural contexts. Group formation is a complex social matter and extension agents are often ill-equipped to do this properly. Niels Roling from Wageningen Agricultural University, whose work on classification of extension systems is cited earliers, emphasizes (on the basis of his field research) that the change in behavior required by encouraging group extension calls for attention to five different elements: mobilization, organization, training, technical and resource support, and replication and maintenance. All five elements have to be addressed if the approach is to be successful, since they are interdependent. It is worth noting that various extension systems use groups and take into account the same principles, albeit to a greater or lesser degree. An interesting example of group extension is the 'groupements villageois' formed under 'animation rurale' programs in West Africa. This approach has been very successfully grafted onto the commodity approach, in, for example a Bank financed cotton and food crop project in Mali Sud through CMDT.

Cooperatives are also examples of group extension, but their level of success is almost inversely proportional to the level of administrative interference from governments. They need to grow in response to felt needs from farmers, not as a result of pressure from outside. Sometimes such groups have fallen into the hands of elites who abuse them for their own ends. Where they are successful, however, they may even reach the stage of funding their own extension agents, who receive training and technical support from government, e.g. the West Highland Rural Development Project funded by the World Bank in Cameroon. Such agencies are likely to have a greater sense of service to their members than many others, and have the advantage of relieving government of the cost of providing extension staff at the grass roots level, and perhaps beyond.

CONCLUSION

A sound agricultural policy framework is needed for every country. Research and extension needs must be defined within the hierarchy of national development priorities, and appropriate budgetary provisions made to enable staff to be used effectively. As a general rule, no more than 70% of recurrent funds should be spent on salaries, leaving 30% to ensure staff mobility and effective operation. Some countries now devote 85-95% of recurrent funds to salaries, so the staff become desk-bound, frustrated and ineffective. Maximum use should be made of private sector initiatives for crops that can bear the cost of an extension or research service; and self-reliant farmer organizations should be fostered and encouraged to take over appropriate government service support functions at the local level, since they are often both costly to maintain, and inefficient.

In the cost context one needs to look closely at extension methodologies. All successful systems emphasize face to face contact, and practical field-based training and supervision. The ratio of extension workers to farmers will vary according to the density of population, settlement patterns, and nature of the farm system, but ratios can be kept down by careful use of farmer groups and conjunctive use of other media, such as radio, visual aids or even television. Too often, radio extension support programs are planned quite separately from ongoing field extension programs. Ideally they should be part of the same program and planned together. This not only has a multiplier effect but can often obviate the need for increasing staff at the field level.

Turning from costs to technology, one cannot over-emphasize the need for close liaison between research and extension to ensure that research programs are relevant to farmers' needs. Among other things this requires that validation programs are effectively and jointly carried out on farmers' fields with active farmer participation. It is also clear that one needs to institutionalize effective procedures for joint review by research, extension and adaptive trial programs. Organizing periodic workshops for research staff with extension subject matter specialists has proven to be a valuable way of promoting communication and understanding.

Any extension program must be well organized and planned with specific objectives and responsibilities for all staff (particularly for subject matter specialists). Extension programs benefit from a single line of command, and continuous, task-focused training programs for agents. Trained manpower at the technical specialist level is a likely candidate for increased investment in many countries as well as training for extension and research managers.

Many policy makers continue to remain woefully ignorant and therefore unsuccessful in their failure to recognize that a very large proportion of the family farm labor force is female and that the role of women in African farming systems is, in many cases, pivotal. This realization calls for some fundamental rethinking of much of the conventional wisdom on crop and livestock production, the ways and means of improving the technology for increasing productivity, and the extension systems for promoting information flows from farmers to researchers and from researchers back to farmers.

Investing in extension should be regarded by both international and national agencies as a continuous learning process. The fourth workshop of an ongoing international series on extension was recently sponsored by the World Bank and other agencies (Ivory Coast, February, 1985). Two have previously been held in Asia. The third was held in Kenya in June of 1984. This illustrates the active and continuing interest displayed by the World Bank in learning the lessons of past experiences and in helping member countries strengthen their technology generation and dissemination capability.

It bears repeating that there is no single blueprint for the best extension approach. Each must be tailored to meet particular conditions. Blind adherence to a successful system elsewhere could be a recipe for disaster. In particular, one must fully understand a country's existing extension system to be able to recommend improvements that can be introduced with a minimal of upheaval.

SUMMARY OF LESSONS FROM EARLIER WORLD BANK WORKSHOPS

We learned from the first Asian Workshop in Thailand that, before launching any additional activities in the extension field, it is absolutely essential to obtain the support and understanding of not only the farming population, but also, and more particularly, the local and central government officials concerned with the administration of the area. As well, if recommended practices involve the utilization of such purchased inputs as fertilizers or seed dressings it is imperative to ensure that supplies are readily available to farmers. If this is not the case, farmers lose confidence and the extension message is regarded as so much rhetoric that cannot be put into effect.

The second Asian Workshop in Indonesia, focused specifically on the linkage between researchers and extensionists. The most important insights from that workshop include:

- There must be increased emphasis upon on-farm research, with special attention to socio-economic studies and feedback from extension staff and farmers;
- This increased emphasis must, if it is to be effective, be accompanied by greater interdisciplinary collaboration among researchers and between the researchers, farmers

and extension staff working in the field, whether at the village or district level, or as subject matter specialists;
- All agricultural institutions, policies and procedures must become more responsive to the interests and needs of the farming population (particularly to the vast majority of small farmers) while simultaneously being responsive to national economic development needs;
- A final, and perhaps key point, is the importance of management. Management (whether research or extension) must be alert, dynamic, responsive to local and national needs, and above all, aware of the interdependence that exists between the farming population, the staff of the extension system, and the researchers who support them. Management should depend to a large extent on farmers and extension staff for guidance as to what needs to be done.

The workshop in Eldoret, Kenya, held in June of 1984 drew attention yet again to the need to guard against simplistic solutions. It emphasized the importance of ensuring that research work is directed toward the real needs of farming populations. The farming systems approach pioneered by Michael Collinson of CIMMYT has much to offer in this respect to Sub-Saharan Africa.

The Kenya workshop raised significant questions as to the cost effectiveness of different extension systems and, in particular, queried the levels of expenditure apparently required to institute and operate an effective Training and Visit extension system in Africa. Finally, it emphasized the role of women in African farming and the need to take this more fully into account than has previously been the case.

AN EXHORTATION

It is vital to speak with candor of perceptions of the strengths and weaknesses of research and extension systems. To be really useful one needs to know the actual situation, what happens (not what is supposed to happen), and then to make judgements as to how applicable the systems might be in particular circumstances.

Finally, it is obvious that there is no panacea for all the problems facing agriculture in Africa - they vary enormously in their nature and in their degree of severity. The important thing to begin to do is to take stock of the situation in the light of perceived national constraints and resource potential, and existing research and extension systems. This stock taking should take full account of the economic, social and cultural conditions of the people (men, women and children) who make up the farming population. Without their support and understanding the systems are valueless. It should then be possible to begin to find ways and means of improving existing systems to make them more effective and efficient.

Chapter Five

SUCCESSFUL AGRICULTURAL EXTENSION: ITS DEPENDENCE
UPON OTHER ASPECTS OF AGRICULTURAL DEVELOPMENT.
THE CASE OF PUBLIC SECTOR EXTENSION IN
NORTH-EAST AFRICA*

Nigel Roberts
The World Bank

INTRODUCTION: THE VULNERABILITY OF EXTENSION
PROGRAMS

Few observers of the agricultural development scene would
dispute that effective extension can play an important role in
the improvement of small-holder agriculture in developing
countries, but equally few would claim that extension offers
any magic solution to the problems of agriculture. These days
most proponents of agricultural extension as a development
strategy sound a note of caution when extolling its virtues.
Take for example the comments of Donald Pickering, Assistant
Director of the World Bank's Agriculture and Rural Development Department (see also his chapter in this volume). In his
keynote address to the West African Extension Workshop held
in February 1985, Pickering pointed out that:

> the benefits to be obtained from improved extension are
> ... closely related to the availability of improved tech-
> nology, and to the supply of inputs, credit and market
> infrastructure. Since the latter are weak in many parts
> of Africa, improvements in extension and research must
> go hand in hand with strengthening the other factors.
> Indeed, in some cases, focus on extension may not
> warrant the first priority [1].

In addition, the 1984 Position Paper of the USDA Extension Service, while arguing the case for extension, is careful
to qualify its enthusiasm:

> Extension provides a unique and important function in
> agricultural development programs. Its effectiveness is

---

*The views and opinions expressed in this chapter do not
necessarily reflect the position or policy of the World Bank
and no official endorsement should be inferred.

(however) dependent upon a reliable source of relevant technology, a dependable supply of agricultural inputs, and favorable governmental pricing policies and markets. Given those supporting factors, extension can provide direct, dependable, objective information upon which farmers can make decisions of benefit to their families and their society. It provides a unique basis upon which the factors of research, inputs and markets can find relevance and acceptance at the place of primary importance - the individual farm [2].

These 'supporting factors', or the context within which the transfer of technical information takes place, is the theme of this paper. It examines four key areas of concern to public sector extension that exist in all countries: the agricultural research network and its links with extension; credit and input supply systems; farmers' incentive structures; and the effective use of government extension staff. It is worth recalling how problems in these areas will limit the impact of extension, since the history of extension development in Third World countries is littered with examples of projects that failed precisely because their designers ignored this context.

This discussion focuses on the Horn of Africa. This is an area where the agrarian support mechanisms, of which extension is a part, are particularly weak. Countries such as Somalia and Ethiopia, therefore, present unique problems and challenges for extension system designers.

THE CASE OF INDIA: A FAVORABLE ENVIRONMENT FOR EXTENSION

The World Bank is best known in the extension field for its close association with the development and propagation of the Training and Visit (T&V) system of extension, first adopted on a major scale in India in the middle 1970s. While it is not my wish to discuss T&V per se, an examination of the Indian experience will help us understand what is needed for successful extension and will therefore provide points of reference when we look at the Horn of Africa.

One important difference between India at that time and the Horn of Africa today is that - by common agreement - extension in India in the 1970s was the weak link in the agricultural service package available to the farmer, at least insofar as those with irrigated land were concerned. Otherwise, India was well-primed for the uptake of new technologies - attractive but undisseminated 'green revolution' research results existed, inputs could be purchased widely, credit was freely available, grain markets worked with tolerable efficiency, and producer prices were renumerative enough to invite farmer investment in improved seeds and fertilizers. India's road network and infrastructural base were also sound. Extension, however, was ineffective. Providing technical advice on agriculture was only one of many functions for which village-level community development workers were responsible, and it was a function easily neglected in favor of other more pressing and (in many an administrator's view) more worthwhile tasks.

Thus, when T&V arrived in India, it faced a situation where extension reform was a high priority on the agricultural agenda, and where the preconditions for successful extension were already in place. The advent of a sound extension management system completed the circle, and a significant impact on production was possible. Examined from this perspective, the recent authoritative study on the results of T&V in Haryana State [3] is not only a vindication of T&V but also a testament to the potential of extension per se under favorable conditions.

THE CASE OF THE HORN OF AFRICA: A HOSTILE ENVIRONMENT FOR EXTENSION

In many respects, North-East African extension services face a set of ambient conditions drastically inferior to those that exist in India.

Inadequate Research Structure, Poor Research-Extension Cooperation

A fundamental problem facing extension in the Horn is the lack of information available from research in a form fit for dissemination. Two important and related reasons for this are the weakness of the research institutions, and the lack of proper cooperation between research and extension organizations.

Building a research capacity is a painstaking process requiring a consistency in goals and management procedures and a continuity in staff service and financing rarely achieved in recent years in Africa. Networks built up prior to independence have deteriorated (e.g. in Sudan), while those with more recent genesis have been subject to fluctuating government and donor commitment (e.g. in Somalia). Jon Moris has written eloquently of the degeneration of East African research networks [4]. The pre-independence structures he describes benefited from a set of circumstances that disappeared with the transition from colonialism - a supply of cheap young expatriates motivated by the prospect of publication for a like-minded international audience; a clear monocrop orientation in the work program; easy access to imported fuel, equipment and spare parts; and the license to dismiss local staff who lacked diligence.

At present, the best African researchers are drawn away from their countries by better financial rewards and professional prospects in the international research network. Those who remain at home belong to civil services that neither recognize excellence nor penalize mediocrity. As well, they find themselves working under shakened and erratic financial circumstances that often make reputable research impossible - a situation that brings with it professional isolation and demoralization. As Moris explains, this is hardly conductive to the production of information useful to farmers, nor to the forging of any meaningful partnership with extension:

> The main output from most African research stations is their annual reports. As likely as not (judging from those I have read in Kenya, Tanzania, and Uganda) these present raw experimental results in uninterpreted statistical form. They will be written in stilted scientific jargon, intelligible only to the writers. The (extension service) will have little capacity to interpret, package and translate such information. Perhaps this is just as well. African research stations until recently did not record labor inputs, assumed access to inputs, and ignored risks. Rarely were results adjusted to account for farmers' varying managerial skills and resource availabilities. Such 'recommendations' arrive eventually at the district office, where some hapless recent graduate struggles to translate them into the simplified vernacular required by the field agents. The contact staff then in turn read out these instructions - which from previous experience they do not trust - public meetings that are attended mainly by the oldest, semi-retired farmers. In such meetings decorum inhibits farmers from voicing their frank disagreement with many of the technical recommendations. And, finally, when they get home they may choose to tell their wives what was said; but often they do not [4].

Much of the impracticality and some of the deficiencies of these weak research organizations might be overcome if research and extension institutions pooled their talents, particularly on research directed towards field adaptation. Serious cooperation is rare, however, and marked structural and attitudinal separation is commonplace. In Ethiopia, for example, the Ministry of Agriculture was, until recently, only one of several user organizations represented on the Board of the national research institution (albeit the major one), and its influence on the definition of research priorities at the macro-level has been weak. At the field level, a few joint adaptive trial sites have been developed, but these hardly constitute the type of dynamic interaction between developer and disseminator of technology that the notion of a research-extension continuum presupposes. Researchers have had no involvement in day-to-day extension operations, and have provided little worthwhile technical advice to extensionists. In addition, there are no local-level liaison groups or joint working parties of any operational significance.

One cannot ascribe all blame in a situation like this to research, however, if the extension system is essentially undynamic and is not disposed to focus in on analyzing and solving farmers' problems, as in the case of Ethiopia, it is difficult to expect research to come forward with the kind of collaboration required to achieve solutions - the pressure generated by farmers' needs must first be channelled through extension. The results of this lack of meaningful integration between research and extension are debilitating to both services. Researchers tend to stray into the exotic and the publishable in their on-station work, and to forego the discipline imposed by a need to derive results which can stand up to the stringent economics of the farmer's world. For research, the price of isolation from extension is a tendency to indulge in the irrelevant. Extension, on the other hand, loses its access to a technical expertise it cannot usually dispose of by itself, and which is essential if it is to offer useful advice over a sustained period.

In both Ethiopia and Somalia, the extension services have tried to compensate for the lack of vigorous research-extension linkages by carrying out their own adaptive trials. At best, this usually involves a costly investment in technical assistance to compensate for a lack of the requisite staff, with poor prospects of continuity. At worst, it results in unscientific procedures and misleading results.

Deficient Input Supply and Credit Mechanisms
Extension in the Horn of Africa also operates in an environment in which the all-important input supply and credit systems are poorly developed. In Ethiopia and Somalia, for example, the provision of such services to farmers is dominated by the public sector. In part, this is due to the subsistence nature of production - cash flow in the rural economy is in many places inadequate to support the growth of commercial services, and the government feels an obligation to step into the gap. In addition, however, most governments' policies have supported the involvement of the public sector, attempting to control or curtail the influence of entrepreneurs and to enlarge the sphere of government interest (a process which is now being reversed in Somalia, but intensified in Ethiopia). The reasons for this vary from ideological belief in the virtues of public sector involvement, to mistrust of the middleman, to the pursuit of personal power and wealth by individuals in government.

79

The results of government control over inputs supply and rural credit, however, are not encouraging. The supply and financing of inputs on a national scale in the Horn of Africa is an activity which would tax the capabilities of the most sophisticated of public sector institutions. It is a task beset by myriad difficulties - by having to project demand from fragmentary field data, by an insufficiency of precise research results to guide in the selection of inputs, by the complexities of inputs procurement on international markets with scarce foreign-exchange resources, and by all of the problems associated with the delivery of merchandise from the port to the farm - enormous distances, poor roads, worn-out transport, lack of fuel, and insufficient storage. It is hardly surprising that farmers suffer from inappropriate inputs often delivered too late, and that repayment rates for seasonal credit lapse in consequence.

One might argue that in such circumstances reliance on monetized inputs should be de-emphasized in favor of improved husbandry, and that extension should stress the best utilization of traditional or existing factors of production. But here again one confronts inappropriate biases in research. Moris believes that:

> (African) research scientists have not comprehended that a technology-intensive program will usually also be organization-intensive. The enormous benefits which seemed to accompany particular innovations on the research farm, made proponents of technology transfer blind to the complexity of the associated support system ... as African nations experience greater economic difficulties, their leaders are taking steps which make the local support network even more unreliable than it was a decade ago. Yet the research scientists continue to produce recommendations that depend upon easy access to insecticides, herbicides, fuel and equipment [4].

The involvement of governments in the provision of inputs and credit impacts directly upon the work programs of their extension services, often the only and always the most numerous source of public sector agricultural personnel in the countryside. Some argue that the involvement of extension personnel in inputs supply and credit provision is inevitable or even desirable. They contend that extension messages are usually linked to the use of inputs and the ability to purchase them, and should thus be seen as an indivisible part of the promotional package, best delivered by a single actor who can comprehend its rationale. One can indeed point to successful commodity-based extension schemes in Africa in which the extension agent provided both technical advice and inputs assistance (for example, the British-American Tobacco schemes in Kenya, or the activities of the West African Cotton Companies like the Compagnie Malienne pour le Developpement de Textiles).

Other individuals, however - and T&V proponents usually join this group - contend that it is damaging to involve extension staff in inputs and credit matters. The main reason cited is the diversion of effort away from the extension function. Other objections include the confusion of roles that can emerge when an extension agent is asked to act as a government regulator as well as a farmers' adviser, and the possibilities for graft that surface if the agent is involved in determining who should obtain access to scarce (and often subsidized) inputs, or in collecting debts. The T&V handbook has this to say:

> Extension personnel ... should not be assigned responsibility for regulatory functions, supply of inputs, and collection of statistics ... such activities, which often have to be performed in the peak agricultural season, when extension staff are most needed by farmers in their fields, will consume much of extension staff's time and divert their attention from their main responsibilities, undermine farmer's trust in them, and interfere with their necessary systematic and timebound plan of work [5].

Extension in the irrigated cotton corporations of Sudan illustrates this argument. There, extension is only one of a number of responsibilities assigned to the corporation field inspector. His other duties include overall management of a designated area of the scheme (planning and supervising the execution of mechanical operations, inputs distribution, etc.) and the inspection of tenancies within that area to ensure that corporation regulations are being followed. Historically, extension and tenant education has received scant attention in corporation agriculture, in comparison to direct management activities - partly as a consequence of the inspector's non-extension responsibilities. In the Sudanese government's own analysis, this neglect of extension is an important reason for the low crop yields. In 1983-4 therefore, the Government was inspired to prepare an inspectorate reform program in which the importance of extension would be assured through a redefinition of the inspectors' responsibilities to exclude such functions as input supply.

Insufficient Production Incentives for Farmers

Extension efforts in the Horn of Africa have, since independence, been more or less severely hampered by a lack of adequate incentives for producers, a factor which works against farmers' willingness to invest in new technologies.

The incentive structure within which a farmer operates has been defined as:

> all those aspects of the farmer's environment which affect his willingness to produce and sell [6].

While this includes such diverse factors as the security of land tenure and the opportunities for investment and consumption, the two most widespread and damaging forms of disincentive in the Horn are an unattractive price structure, and inefficient markets.

The link between adequate output prices and yields has often been alluded to. The World Bank's Accelerated Development in Sub-Saharan Africa, for example notes:

> It is now widely agreed that insufficient price incentives for agricultural producers are an important factor behind the disappointing growth of African agriculture. The importance of price policy comes out strongly in project experiences. A recent review of 27 agricultural projects undertaken by the World Bank noted "the almost overriding importance of producer prices in affecting production outcome and production levels, often cutting across the quality of technical packages and extension services. Seven out of nine projects implemented under favorable prices achieved or surpassed their production objectives; 13 of the 18 under unfavorable prices failed to do so". This idea is also borne out strongly in microlevel studies, which indicate substantial farmer responsiveness to price [6].

Of particular importance to extension, given the previously mentioned reliance on input-intensive recommendations, is the benefit-to-cost (BC) price ratio for fertilizer - i.e. the cash value of the incremental production generated by the application of fertilizer, as compared to the cost of the fertilizer itself (with fertilizer commonly constituting the most important and most costly element in the input package). In reasonably reliable rainfed conditions one would not expect farmers to show sustained interest in purchasing fertilizers if the BC ratio were less than 2:1. All too commonly, it is.

Ethiopia provides a telling illustration of the impact of deteriorating fertilizer BC ratios on production and of the implications of this for extension. Throughout the late 1970s, the price relationship between di-ammonium phosphate (DAP, the fertilizer of choice) and farmgate grain prices remained fairly favorable, and fertilizer consumption increased from 10,000 tons in 1975/6 to 49,000 tons in 1979/80. One would like to be able to ascribe this phenomenon primarily to extension, since fertilizer use was the principle extension message.

That it cannot be thus attributed appears to be confirmed by developments in the succeeding period.

In the early 1980s, fertilizer prices increased rapidly while farmgate grain procurement prices remained more or less constant. Between 1979/80 and 1981/2, fertilizer prices doubled, and annual fertilizer consumption fell dramatically by almost 40% to 30,000 tons. With an average yield response of between 4 and 5 quintals of grain to one of DAP, this represents a loss of production potential of some 800,000 tons of grain (which would have added 15% to production in 1981/2).

In the Ethiopian situation, one cannot explain this deterioration in incentives as the operation of free market forces. Government intervention was a crucial factor on both sides of the equation, including expensive international fertilizer procurement decisions and imposed grain price controls reinforced by the expansion of compulsory quota deliveries to state marketing channels.

Marketing inefficiencies can be ascribed in part to over-zealous government involvement, particularly in export-crop marketing. Even with an efficient marketing operation, the great distances and difficulties of access in the Horn of Africa make marketing a high cost business. Government-dominated cooperatives, however, or parastatals such as Ethiopia's Agricultural Marketing Corporation or Somalia's Agricultural Development Corporation tend to exhibit serious inefficiencies endemic to civil service enterprises: over-staffing, a lack of cost-consciousness, inadequate operating budget allocations, and a lack of qualified managers. In addition, they often hold a monopoly in their field and are not subject to the rigors of competition. Much of the margin for these inefficiencies must be financed by the producer, who is paid less than he might otherwise be, and often less than an attractive price. In addition, parastatals often pay producers long after purchase, and may periodically be unable to collect or store their output. The cumulative effect of these shortcomings is depressed production, as farmers settle for levels of output commensurate with marketing risk - however fine the quality of extension advice may have been. The Sub-Saharan Report summarized it thus:

> ... the crop marketing agencies are a major point of contract between peasants, the money economy and the state bureaucracy. Unless the marketing transactions are done fairly and efficiently, there are high risks of peasant disaffection from both the bureaucracy and the market economy [6].

Impediments to Effective Use of Extension Staff

The final set of constraints to effective extension relates ultimately to shortcomings in the framework of governance within which extension staff must operate. One could include several factors under this heading, such as poor infrastructure and the resultant strain on the capacity of extension managers to coordinate operations and exercise adequate supervision, as well as the intensifying economic decline in the region as manifested in scarcer and scarcer access to foreign exchange (required not only for agricultural inputs and equipment for farmers, but also for the vehicles, spares and fuel vital to extension's mobility in the field). This discussion, however, will focus on problems associated with the management of extension's most precious resource: the staff of the service.

Production ministries in North-east Africa in the 1980s face severe budgetary restrictions, and national employment policies which (at least until recently) have guaranteed employment to graduates from secondary and higher education institutes. The result is a civil service disproportionate to the funds available to finance its activities, and a salary bill which consumes most of what is available for recurrent expenditures. This has two equally deleterious effects. First, salary levels do not approximate what is required to motivate public servants, or even to keep them on the job. In Uganda, the size of the civil service in relation to available budget was such in 1982/2 that the monthly salary paid an extension agent was estimated by the World Bank to be sufficient for less than a week's food for an average-size family. Consequently extension agents stayed at home farming. Second, funds remaining for operating expenditures are woefully inadequate and housing and travel allowances, fuel, motor vehicle maintenance and farm demonstrations are grossly neglected in consequence. In 1982/3 in Ethiopia about three-quarters of the recurrent budget of the Ministry of Agriculture was used for salaries. The operating budget of the Sudanese National Extension Administration (NEA) remained static between 1980/1 and 1983/4, despite increasing maintenance prices and requirements. As a result, spare parts purchase and plant maintenance became impossible after the first quarter of a financial year, and NEA's important printing presses and photographic labs lay idle much of the time. A number of important documentary films and color slides shot in that period are unexposed to this day. Worse still, average fuel allocations of 5-10 gallons per vehicle per week were insufficient to permit NEA any kind of meaningful field program. Obviously extension cannot function effectively in such situations.

A lack of adequate financial reward and a paralyzed working environment are deeply discouraging to those who are professionally trained, particularly those trained abroad. An understandable consequence is the drift of these scarce and valuable personnel out of government service, often to jobs abroad (most notably to the Gulf countries in recent years).

The civil services that they leave behind are demoralized institutions with little esprit de corps. In such an atmosphere a donor project - usually of short-duration, and often lax in its financial controls - may be viewed by its managers as much for the opportunity it offers for personal and illicit gain as for its development potential. This adds little to the long-term health of government institutions.

There will always be capable staff at all levels who persist in their efforts to do a worthwhile job, despite the financial and managerial shortcomings that rule their lives. In the field, however, MOA extension staff also battle with severe organizational uncertainties beyond the control of their parent ministry. The first, alluded to previously, is the tendency of government officials to load extension staff with the responsibility for inputs and credit work; with data collection and cooperative administration; and even with non-agricultural rural development functions such as overseeing the provision of water supplies or the maintenance of rural roads. And secondly, there is the spectre of insecurity in areas within each major country in the Horn of Africa. At various times in the 1980s this has made organized extension difficult or impossible in the Bugundan areas of Uganda, in Southern Sudan, in the Wollo Tigray Regions in Ethiopia, and in North-West Somalia.

CONCLUSION: THE DESIGN OF EXTENSION PROGRAMS IN A CONSTRAINED ENVIRONMENT

Those of us in bilateral and international aid institutions who specialize in technology-transfer matters often become too involved in debating the merits of one extension approach over another, and lose sight of the overriding influence of the context in which extension operates. Of course it is important, when establishing a new system, to employ a rational methodology. But it is not enough merely to erect a promising extension structure and hedge it against short-term financial adversity with special project funds. There will be little meaningful impact on production if the research system is defunct, if access to inputs and credit is inadequate, if pricing and marketing structures work against higher production, or if extension staff face impossible working conditions once the project period is over.

Apart from being aware of the vulnerability of extension systems to their environment, what can a project designer do to ensure that extension initiatives are not buried under the weight of problems? The initial and obvious step is to assess whether factors exist, beyond the scope of influence of an extension project, which are intolerably hostile to extension development. This is not a simple decision. First, one must judge the extent to which a new extension program can in fact influence the ambient circumstances, or at least bulwark itself against them for a period of time while their reform takes place.

Clearly, there is little profit in establishing an extension program in areas in which civil servants' lives are in danger from hostilities. But what do you advise, for example, when fertilizer benefit-to-cost ratios are structured against the interests of producers if extension can offer cost-saving messages that will (at least temporarily) render fertilizer use attractive? Or, what if there is no commitment in government to focusing research on farmers' problems, but extension reform could, perhaps, create sufficient pressure on research to achieve its re-orientation? Under such circumstances, will it help or will it hinder the nations's agricultural development if a new extension program is introduced? Should you wait until the conditions favor extension - until a viable research network is in place, until civil service compensation has been restructured, until adequate incentives to farmers pertain? Or should you accept major imperfections in the anticipation that an impressive extension project will exert a beneficial influence on the sector as a whole?

Naturally, one's judgement will depend upon the dynamics of a particular situation. Yet there are some pragmatic guidelines which can help the designer approach these complex questions.

First, it is inappropriate to focus on extension if there are serious deficiencies in the research system, or if there is little prospect of a useful supply of recommendations becoming available to extension in the near future. Normally these two problems occur together, with the second resulting from the first. An extension network which is reasonably well-funded and staffed and tolerably well-managed can usually benefit from research-extension linking initiatives designed to produce information useful to extension within two or three seasons (linkage arrangements such as joint farming systems research and on-farm trials work).

To reform extension in the absence of a credible research effort, or even in the hope that a fundamentally weak and unproductive research network can be turned around in short order by the pressure generated through an improved extension effort is to adopt a gambler's strategy. Rather than risking a situation where the extension service, however well structured and managed, comes to farmers with advice that is irrelevant or even harmful, it is better to work first on reforming research. There is more than one country in the region that has never known an effective extension service; to wait another three or four years for the start of a viable program is preferable to discrediting extension for another decade.

Similarly, it benefits no one to push for major invest-ments in extension if farmers are not motivated to adopt the primary technologies. Inadequate incentives for farmers are normally related to deliberate government policies, often policies designed to transfer revenue from the agricultural to the urban community. As such, these policies have major political significance and will not be susceptible to alteration to help a newly-reformed extension service achieve success. Extension reform should not be attempted unless producer incentive structures make sense.

Third, there is probably little to be gained by significant extension investments if the conditions of staff employment make full-time effort irrational, and if public service regulations prohibit the introduction of a package of benefits that could alter this situation.

A realistically designed extension project can, however, overcome many constraints and can indeed work if the three factors just discussed are marginally, rather than grossly, adverse. For example, an inadequate road network, large distances, and a limited capacity to service and repair motor vehicles can be tackled by restricting extension services to those areas with better roads, and within range of district supervisors' offices (an approach employed with success in Ethiopia in the 1960s). Staff who are poorly but not intolerably poorly paid can probably be motivated to work hard if they are provided with a judicious mix of benefits comprising higher allowances, free housing, personal transport, and attractive scholarship opportunities. A mix of this type, which does not increase salary benefits per se, may well be acceptable to the country's civil service commissioners.

If there is no immediate alternative to the involvement of extension staff in inputs distribution, this too can be reconciled with the requirement that extension staff focus on extension by carefully programming the time allocated to it. Two days in six might be given to input and credit matters, or better still, two staff in six might be assigned the responsibility. The Sudanese extension reform has stripped administrative and regulatory functions from field staff and systematically divided their time between extension and direct field management. In the research sphere much can be done to improve the flow and relevance of technologies by instituting local-level working links. In Ethiopia, research-administered coordinators work at local stations, organizing regular extension training by research, joint field tours, on-farm trials, and joint pre-seasonal workshops to determine extension recommendations.

In closing, let us review two contrasting examples from the World Bank's extension portfolio in North-East Africa. They reflect similar judgements by my colleagues as to when it is or is not profitable to make significant extension investments. In Ethiopia a pilot T&V project has been functioning since 1983, and has been singularly well-managed. On a local level, it has had a significant impact both on crop production and on farming practices (popularizing on-farm soil conservation, and the stall-feeding of cattle, for example). It has also induced a moderately efficient research organization to gear itself much more convincingly towards farmers' concerns. Yet, at this time, the World Bank has no wish to support extension on a national basis, since it is apparent that the Government is committed to an agricultural pricing and marketing policy that does not encourage farmers to invest in the increased levels of fertilizer use necessary to raise cereal production in a meaningful way. In Kenya, however, a 1981 T&V pilot was swiftly followed in 1982 by financing for a national project, as the World Bank felt that the essential elements required to raise crop production were either in place, or would come into place in time. In this case, extension has been used to spearhead a revitalization of the agricultural sector. In one case delay, in the other case, movement; in each case, the Bank's extension strategy was determined through an assessment of the environment in which extension must operate.

NOTES

1. Pickering, D.C. (1985) An Overview of Agricultural Extension and Its Linkages with Agricultural Research: The World Bank Experience. (Paper delivered to Workshop on Agricultural Extension and its Link with Research in Rural Development). Yamoussoukro, Ivory Coast.
2. Watts, L., and the International Task Force of the Extension Committee on Organization and Policy (1985). New Directions: The International Mission of the Cooperative Extension Service - A Statement of Policy. Washington, DC: US Department of Agriculture.
3. Feder, G., Lau, L.J., Slade, R.H. (1985) 'The Impact of Agricultural Extension: A Case Study of the Training and Visit Method (T&V) in Haryana, India'. Washington, DC: World Bank.
4. Moris, J.R. (1983) Reforming Agricultural Extension and Research Services in Africa. (Discussion Paper II). London: Overseas Development Institute.
5. Benor, D., Harrison, J.Q. and Baxter, M. (1984) 'Agricultural Extension. The Training and Visit System'. Washington, DC: World Bank.
6. Berg, E. et al. (1981) (African Strategy Review Group), Accelerated Development in Sub-Saharan Africa -- An Agenda for Action. Washington, DC: World Bank.

Chapter Six

MAKING EXTENSION EFFECTIVE:
THE ROLE OF EXTENSION/RESEARCH LINKAGES

J. Kenneth McDermott
University of Florida

Conventional wisdom holds that the US Extension Model does not work in the less developed countries (LDCs). The implication is that US personnel knew only the Land-Grant model and attempted to transfer it lock, stock, and barrel in the technical assistance program.

The contention of this discussion is just exactly the opposite. Extension is something like someone once said of Christianity. The 'traditional model' has not worked because no one has really tried it. It is not so much that the model did not work, as it is that we took only bits and pieces of it overseas. What we took had some resemblance to our own system, but it had some vital pieces missing. We did not even take the form overseas - only fragments of the function and the label, not concept but label.

The US Extension Model if it is a 'Model', is one of the richest models of any kind in the world. Over the history of extension, now approaching 100 years, it is difficult to think of much that has not been tried in the US system. Thus, whatever model that would have developed would likely have something about it that looked like the US model. So much has been tried, for example, that there are probably many 'US Models'. The T&V system made famous by the World Bank has very specific reflections in the US system. Extension also has significant attributes for development that go beyond technology. It is clear that we cannot discuss the total system or our total experience in dealing with it in this seminar. It is also clear that the 'entire model' could not have been taken overseas.

This paper focuses on two major problems that will lead to comments on other aspects. The two problems identified from the so-called US model are <u>inadequate logistical support</u> <u>and inadequate technical support of the local agent.</u>

First, we cannot deal adequately in the LDC context with either research or extension alone - we must deal with the total Technology Innovation Process. The entities involved with research and extension are a system, and the performance and welfare of each is so closely bound to the other that treating them separately is an error, perhaps a fatal error. Further, the performance of this research and extension system is accurately measured by the performance of the local extension agent - the prime interface between the technology innovation system and the farmer.

The problem of logistical support can be dealt with quickly. There are simply more agents than can be supported. In recent years, at least, extension has been an employment agency more than a development agency. And in almost no case is there significant local support for extension.

The problem of technical support is more complex. In the United States a local agent will receive more in-service training in one year than most LDC agents will get in their careers. It is not uncommon to see an LDC agent refer to class notes made ten years or more before in an effort to help farmers. Often the only instructional material available is a commercial poster.

Our reasoning with respect to technical support is based on the Technology Innovation Process (TIP) Model and some derivatives of it. (See Appendix A for full discussion of the TIP model.) In the current LDC situation, almost always research stops too soon, extension starts too late, and a fatal gap is created in the technology innovation process.

We must see technology innovation as a publicly supported effort to improve agricultural production in the public interest. That means that the only way success of either research or extension can be measured in the typical LDC is by the innovations in agricultural production adopted by the farmer client. The development of an innovation by research that is not put to use simply does not count.

If research and extension must serve society through the farmer, then research and extension must be in close interaction with the farmer. The farmer is critical in agricultural development - the only one who will achieve increases in agricultural production.

Figure 6.1 helps understand the rationale for the 'Fatal Gap' assertion. In even the best of cases, research often stops about midway through the testing process. Testing is not finished until it is done in the systems in which the technology is expected to perform. At the other end of the continuum, extension does not expect to start until the dissemination function. The seriousness of the gap is apparent.

Farming Systems Research is providing an exceptionally effective means by which research can move into that gap from its end of the process and effect the interaction with farmers. It operates in the center of the Technology Innovation Process, getting to know and understand the farmer and his system of farming and testing proposed innovations in those systems in which they expected to perform, and by criteria of those systems.

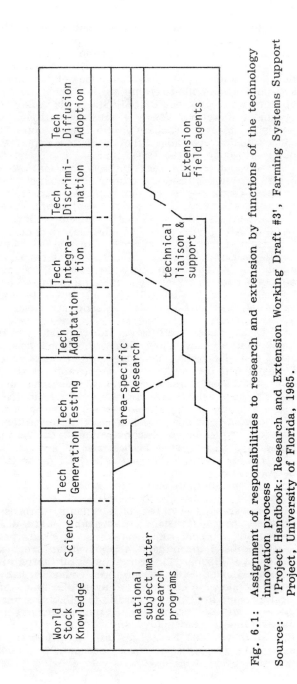

Fig. 6.1: Assignment of responsibilities to research and extension by functions of the technology innovation process

Source: 'Project Handbook: Research and Extension Working Draft #3', Farming Systems Support Project, University of Florida, 1985.

91

As of now, extension has not made a significant move into the gap from its end of the process. Until it does the effectiveness of FSR, indeed technology innovation in general, will be limited. Extension is in many cases a large force of field personnel, with little sense of mission, waiting for someone to give it something to do. The fatal fault is the lack of the what in the United States is called the extension specialist function. Many extension entities have so called 'subject matter specialists', but almost none have developed the extension specialist function to the extent to which it has been developed in the US system.

A major reason for this may be that we in the United States do not ourselves really understand this part of our system. Certainly we do not appreciate it. The best evidence for that assertion is to count the number of extension education courses that deal with the specialist. Jack Claar of INTERPAKS (University of Illinois) helped coin the term 'technical liaison and support' because so many of our own people working in overseas development simply do not understand the 'extension specialist' concept. 'Technical liaison and support' does not really capture the richness of the extension specialist role in the US system, but it does describe two essential functions of that role and may lead to some of the other functions.

If Extension can move 'to the left' in the technology innovation process, then there is a good chance that Extension and Research can develop effective linkages. These linkages will be functions in the center of the process - the lines that assign responsibility to research and extension in the diagrams must slant. In other words, in the center of the process there is no way to clearly distinguish research and extension. A field trial of a worthy innovation is a demonstration - just as a demonstration is literally a field trial, especially if the system learns from it.

For extension to move to the left in the process, it must become more dynamic. It cannot remain a passive entity waiting for another entity to energize it. It must develop a capacity in technology that will enable it to deal effectively with research. The extension specialist is an essential element in energizing extension - extension must add the technical liaison and support function.

Technical liaison and support has three major responsibilities to discharge through several activities. One is to maintain liaison with research in order to know the current best technology alternatives available and about promising alternatives that are becoming available. It must not only know about the technology, it must also have the capacity to understand the technology and to work with it. This requires that perhaps half the TLS staff have formal training to the same level as area research personnel, probably the M.S. degree, and that all have adequate short-term training.

Collaboration with field research in testing and adaptation is the single most effective way for extension to inform itself of technology development. It can give the technology a better test than can research without this collaboration. Finally, for the technology that does stand the test, the extension process is off to an early start. Collaboration also facilitates extension participation in problem identification and problem definition and in deciding what passes the tests.

The second responsibility of technical liaison and support personnel is to establish linkages with input suppliers to improve the chances that the right inputs will be available for that which is technology embodied in inputs. In the case of seed, this unit could recruit producers of improved varieties.

The third responsibility is to provide technical support to field staff. The field staff, by its very posting, will quickly become isolated from the rest of the system, if the system is not energetic in keeping it integrated.

Technical support activities include training of field agents; preparation of reference materials and training aids; trouble shooting and response to agents' requests for help. Training of field agents needs to be integral to the extension program, not an ad hoc service from other entities. Training is the principle means by which information, extension's stock-in-trade, flows through the system. Field agent training needs to be part of the program of technical liaison and support personnel and should be written into job descriptions.

If, in the center of the function, the lines dividing research and extension did slant, many positive things could happen. One is that either entity could compensate to a certain extent for the other. Extension can do much of the FSR type of work. I am convinced it did in the United States. Or, on the other hand, research can compensate in part for extension inadequacy. Either one, operating in this area, can serve to help improve the other.

If both research and extension deal with some of the same functions, it creates a certain redundancy. Some may call it duplication. Whatever it is called, it does not violate any rules of administration. There is plenty of work for both to do, and redundancy is a respected administrative technique for insuring against breakdown in the process.

In my 25 years of working with LDCs, I am convinced that the US system has much to offer to the world. I am further convinced that we cannot free ourselves from our own tradition - that the best way to protect against provincialism of tradition is to understand that tradition as thoroughly as we can.

APPENDIX A*
Technology Innovation Process (TIP) Model

The Technology Innovation Process Model is an over-simplified conceptualization of a process that is more complex and exact than is generally recognized. As with any conceptual model, it does not intend to represent reality. It is presented as an aid in understanding and working with reality. It should accomplish three purposes:

1. to help understand and explain the process with which research and extension must deal,
2. is to stimulate the imagination and help gain insights in managing research and extension,
3. to help facilitate communications among all of the different persons and professions involved in designing and sustaining a research and extension effort.

Technology Innovation is defined as an improved technology in general use by farmers. Unless an 'improved technology' is put into the production process on a fairly broad scale, it is not an effective innovation in terms of the industry and of agricultural development.

I. The Model
The model has eight components, commonly called functions. (See Figure 6.2). It appears here as a simple linear process, although in practice that is seldom the case. The model makes conceptual distinctions between functions that may be difficult to identify in practice. It is not necessary to distinguish among the functions in practice, and in fact it may be harmful to try to do so.

1. The World Stock of Knowledge is held in the International Agricultural Research Centers and in research and extension organizations of other countries. There is not a formal network with coordination and management, but there is networking activity among some of the entities who hold science and technology knowledge. The World Stock of Knowledge includes folk wisdom and traditional technology as well as scientific knowledge and advanced technology. Some of it
...

---

*From 'Project Handbook: Research and Extension, Working Draft #3', Farming Systems Support Project, University of Florida.

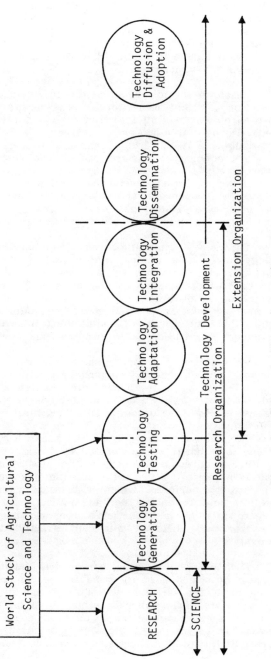

Fig. 6.2: The technology innovation process

is embodied in products - seed, chemicals, implements - some in manuals and books; and some in the minds, intuitions, and traditions of people. Much of it is present in-country. Any country can take advantage of this stock. To a large extent, LDCs do not have to catch UP to the world's technology; they can catch ON to it.

2. Research in this model refers to science, in contrast to technology. Scientific research seeks new knowledge, and it does so by abstracting from the real world. It seeks as much control over variables as is feasible. It is analytical. New knowledge, of itself, has no value to farmers, until it is put into a technology. Farmers cannot use science. They need technology
    However, most technology advances are based on science, and science is the basis for so-called breakthroughs. Technological advance is often stopped for want of new knowledge that only science can provide.

3. Technology generation puts together knowledge, technology, even folk wisdom into a form that serves a useful function. This form may be a commodity, such as seed, or it may be a practice, such as placement of fertilizer. Technology generation synthesizes. It makes new knowledge useful. Technology must serve in uncontrolled conditions and is more useful the wider range of conditions it tolerates. The role of technology generation is to produce new technology alternatives.
    While there is a conceptual distinction between scientific research and technology generation, they often blend into each other in practice. They both use the scientific method, and both can make use of a high degree of training and creativity. Both are essential to agricultural progress.

4. Technology testing moves the technology from the conditions in which it was generated to determine its performance in other conditions. Eventually the new technology must be tested on farms - i.e. in the farming systems in which it is expected to perform. On-farm testing is essential, and if research and extension do not do it, then the farmer will have to do it himself. Farmer testing may be effective, but it will also be inefficient and will greatly delay technology innovation.

5. Technology adaptation serves two functions. It is the process by which a newly generated technology can be fine-tuned to fit the farming system for which it was intended to determine its performance in other conditions. Eventually the new technology must be tested on farms - i.e. in the farming systems in which it is expected to perform. On-farm testing is essential, and if research and extension do not do it, then

the farmer will have to do it himself. Farmer testing may be effective, but it will also be inefficient and will greatly delay technology innovation.

6.   Technology integration fits a new technology into current farming systems. It has three dimensions:

a.   One pertains directly to the system of production. Integration is facilitated by a knowledge of the farmer client and is also facilitated by research on related problems and by extension instruction to farmers on its use. As with testing, integration is essential. The farmer must do it. If he has to do it without research and extension help, it will be inefficient and slow.

b.   A second dimension is integration with the market, both input and product. Much agricultural technology is embodied in a commodity. If that commodity is not available and cannot be made available, a new technology cannot be adapted, no matter what its merit. Integration involves market action to make inputs available or research-extension activity adapted to the lack of input. On the product side, if there is inadequate market, farmers cannot integrate the technology into their systems of production.

c.   The third dimension is integration with national policies. National policy often works through product and input markets and sets conditions the farmer must adapt to. These conditions affect the ways he can deal with new technology. If policies are not adequate and cannot be changed, the conditions they create must be adapted to.

7.   Technology dissemination involves informing farmers of the new technology and helping them figure out how to fit it into their systems of farming.

For simple technology, informing is all that is needed, and farmers themselves can fit it into their systems. Dissemination means 'to seed', and for simple technology, 'seeding' is all that is needed.

The extension demonstration is one of the most effective seeding devices. It may not be as much a 'demonstration' as it is a means by which the farmer's own experimental process is facilitated. Most farmers are both experimental and skeptical. They will not adopt a practice until they have either experimented with it in their own system or have seen it perform in a system almost like theirs. The demonstration facilitates this process and is literally an 'on-farm trial'.

As technology becomes more complex, more assistance is needed from extension to help farmers fit it into their systems.

8. Diffusion and adoption are largely a function of the farmer dynamic. Farmers themselves, through their kinship groups and other social systems, constitute a powerful force, working either to facilitate or to impede diffusion. This farmer dynamic has been responsible for much diffusion throughout history, unaided by research and extension. Extension is most effective when it takes advantage of and encourages the farmer dynamic.

Diffusion and dissemination are distinguished here to reflect the distinction between outside forces and the farmers' own force in the diffusion function of the process.

II. Some Implications of the TIP model

1. Technology innovation is a 'natural' or autonomous process that has been going on throughout history, driven by an innate human desire to improve things. Research and extension have been organized to accelerate the process, not to replace it. Research and extension will likely function best if they understand the process and collaborate with it.

No part of the process can be ignored. If research and extension (or other mechanisms for accelerating innovation) ignore a function, then it will have to be accomplished by farmers themselves - and the process will be delayed, at best.

2. The model puts Farming Systems Research and Extension in context. FSR/D deals specifically with testing in the farming system, adaptation to the system and integration into it. It is through these functions that research and extension begin to come to terms with the farmer and to take advantage of the farmer dynamic. If the R/E system does not address these functions, then farmers are on their own.

3. The TIP model presents no clear line by which research and extension can be separated. As technology becomes 'tested and adapted', the 'on-farm' trial becomes virtually a 'demonstration', and as 'demonstrations' turn up new data on performance of the technology or even confirm old data over a wide area and several years, they are 'on-farm' trials. Thus, the research process shades into the extension process. Extension is probably more effective when it is helping farmers solve their technology problems than when it is merely instructing them from what it knows.

4. The TIP model implies that a country can rely to a large extent on the international technology network for science and

new technology alternatives. It implies even more strongly that the international technology network has little to contribute from the testing function onward through the process.

5. The model also shows that FSR/E probably has reduced potential if left completely on its own. In other words it is heavily dependent on the processes of technology generation and science, just as science and technology must depend on it for the fruition of their efforts. FSR/E completes the research process and initiates the extension process, giving extension a tested farm ready technology. FSR/E also has the potential for sending signals about needs to the technology generation function. Thus, FSR/E may have its greatest value in its capacity to condition the entire technology innovation process, perhaps greater than its own direct contribution. Management needs to reflect this.

II. PRACTICES

Chapter Seven

THE DIFFERENT SYSTEMS OF AGRICULTURAL
EXTENSION EDUCATION WITH SPECIAL
ATTENTION TO ASIA AND AFRICA*

George H. Axinn
Michigan State University

INTRODUCTION

This paper describes current patterns and new trends in
agricultural extension. It emphasizes their effectiveness in
diffusion of new agricultural technology and their applicability
to small farmers in developing countries, with special focus on
Asia and Africa.

To a person who has been struggling with these matters
for over 40 years - first as a young farmer on a small farm
which was declared economically not viable; later as an agricultural extension worker at many different levels in several
states of the USA; still later as a scholar and researcher
trying to understand the categories of patterns, the validity
and reliability of measures of effectiveness, the diffusion of
technology, and the neutrality of scale; and then in more
than two decades of practical field struggle to improve the
human condition, first in Africa, and more recently in various
parts of Asia - the assignment is overwhelming.

In so doing, I am reminded of a line in the book Walden,
written by Henry David Thoreau - an American philosopher
much influenced by his studies of Hindu scripture and Hindu
culture. Thoreau said: 'There are thousands hacking away at
the branches of evil for every one who is cutting at the
roots'. What follows is an attempt to identify and expose the
roots, however well hidden they may be by the branches.

The systems of agricultural extension are many. Every
nation state has one, many have more than one. The numbers
of professional personnel involved are legion. A recent survey

---

*Based on a paper first presented at the Symposium on
Education for Agriculture, 12 to 16 November 1984, at the
International Rice Research Institute, Los Banos, Laguna,
Philippines.

by Swanson and Rassi (1981) [1] shows over 290 000 men and women, working throughout the world in agricultural extension.

The scope of the effort is enormous. It includes indigenous learning systems, which are everywhere, and carry the main burden of agricultural education for many rural people. It also includes exogenous learning systems, sometimes with massive bureaucracies which have been introduced relatively recently, and are struggling with difficult problems of size, management, personnel, program development, and implementation.

But the potential of the effort is significant. The significance of agricultural extension must always be one of its major tests. What difference does it make? Are rural people better or worse off because of agricultural extension? Or, perhaps as disastrous, does it make any difference at all whether or not the world has agricultural extension education systems?

Let us review briefly some of the examples of successes and failures, analyze the types of systems and current patterns of agricultural extension, and then point up some of the issues which are current.

CRITERIA FOR SUCCESS

One criterion which is perhaps most often used as a measure of success for agricultural extension is that of production. If the farmers produce more, agricultural extension may receive the credit, at least in part.

Effective agricultural extension work in Pakistan certainly contributed, two decades ago, to the rapid spread of Mexi-Pak wheat, and close behind it the new short-stemmed rice varieties from the International Rice Research Institute in the Philippines. And on both sides of the Punjab, where new technology fit, where the inputs were made available in a timely fashion, and where the prices of the surpluses which farmers now could sell on the market were high enough to make production profitable, production certainly increased.

Even more dramatic stories can be told of the specialized agricultural extension organizations which work with rubber in Malaysia or tobacco in Bangladesh. In both cases, a relatively small, well-staffed agricultural extension group, under the same management as those who supply the inputs and purchase the outputs from farmers, were able to demonstrate significant success in short periods of time. Bangladesh went from a tobacco importer, with almost no local production, to an international exporter in just a few years, with assistance from effective agricultural extension.

In addition, there are many examples of small-scale local efforts in agricultural extension which have been highly successful. The Small Farmers Development Projects in Nepal are typical of others throughout that part of the world where a very limited number of skilful, committed, disciplined agricultural extension staff have worked closely with small groups of local people and have gone beyond merely increasing production. In some cases, and at least for a limited period of time, they have contributed significantly to the enrichment of rural life, and to the improvement of the human condition among the rural people of their areas [2].

Additional demonstrations of small scale projects may be found in Nepal Hill Areas Education Program and in Khit Phen in Thailand, in the thousands of brigades in rural communes in China, in the village groups of the Semal Udong movement in Korea, in small rural bank branches with successful farm credit programs in the Philippines, and the Ghandi Grams of India. They are living examples that small is beautiful, even in agricultural extension [3].

But there is also the other side of the coin - the examples of agricultural extension education efforts which have not been so successful. Sometimes production does not increase. In some cases, yield per hectare even goes down, and agricultural extension seems to receive the blame. One way out is to suggest that productivity is in the hands of larger powers - if the monsoon fails, if the floods come, or if insects plague in uncontrollable numbers, what can the extension staff do? Or if the targets are set by central government, and are not in the interest of local people, why blame the extension officer when farmers do not adopt a particular practice? If the technology fits large farms, but simply is not feasible for small farmers, how can the extension staff achieve equity goals, or serve small farmers?

Failures often involve a large staff of field workers who are controlled and supported by central government and who suffer from conflicting agendas between local rural people and their superior officers at district or regional level [4]. Unfortunately, the first line agricultural extension agents tend to be poorly trained, are younger than their target farm family decision makers, have very little in common with farm families, and consequently do not have much impact on them. If the extension agent is not given either house or office, he or she may have to find a wealthy rural family to provide a room in exchange for either teaching the children their school lessons or providing special extension and other agricultural services for the owner. When the goals of the landlord are different from those of the agricultural extension administrators, the field agent has a serious problem [5].

Failures sometimes result from excessive influence of foreign sources attempting to assist by bringing in agricultural technology from some other part of the world. Large tractors fail on small farms; submersible electric water pumps for irrigation fail where electricity or the diesel fuel with which to make it are unavailable. And new varieties which require mineral fertilizer and irrigation water fail where chemical fertilizer is too expensive or where water simply is not available when needed.

EXTERNAL AND INTERNAL FACTORS FOR SUCCESS

Some characteristics of success are external to the agricultural extension organization, while others are internal, or characteristic of the organization itself. The first group tends to include policy matters, while the second tends to involve strategic considerations [6].

On the external side, sometimes there is a convergence of agricultural policy and agricultural extension goals, but sometimes there is not. For example, if national price policy makes it profitable for farmers to produce the crop which extension is recommending, the chance of extension achieving its goal is great. However, if national price policy discourages farmers from adopting that crop, extension may fail, no matter how many demonstrations are done, how well radio programs are coordinated with farmer discussion groups, or how well other extension teaching techniques are executed.

This conflict between policy and agricultural extension is well illustrated by Niels Roling when he stated: 'There is persistent evidence that agricultural extension in the developing world is not reaching the poorer farmers and that extension and other agencies tend to focus instead on the better-off farmers who probably represent not more than 20 per cent of the total'. He goes on to explain,

> The objective is often not so much the welfare of the farmers as it is the creation of a surplus for national development, which has often been equated with urban elite development. In the short run, such a policy is much easier to implement with a few larger farmers than with thousands of small ones [7].

Another external consideration may be the technology itself. As mentioned above, if the technology is not likely to benefit the farmers themselves, they will probably not use it. This has been a serious problem in Asia and Africa. The successful agricultural technologies of Europe and North America have tended to feature large scale, capital intensive innovations which fit in situations with a surplus of land, a shortage of labor, and plenty of capital. Unfortunately, attempts have been made to transfer those technologies to places where there is a shortage of land, a surplus of labor, and very little capital. In those situations the technology transfer has obviously tended to fail! [8]

Sometimes extension program plans have called for introduction of a technology where the required inputs simply were not available. An agricultural extension officer is doomed to failure if he or she must try to convince farm families to use an insecticide, for example, when no insecticide is available either at the local bazaar or market, or even at the nearest major market town.

Failure is also predictable when the marketing and distribution system is not there to buy products from farmers which extension officers are encouraging them to grow. Successful attempts to encourage hill farmers in the Himalayas to grow apples have been frustrated when it was later discovered that the cost of transportation to the nearest markets was greater than the market value of the apples.

In addition to external constraints on extension systems, there are also internal constraints. Of those discussed here some relate to personnel management, some to program.

With respect to personnel, it has been very difficult for most agricultural extension systems to employ staff whose social distance from their clientele groups was small enough so that they could communicate effectively. Young workers from urban families may find it difficult to understand farm families, and fail to communicate. Sending male agricultural officers to explain agricultural practices in situations where most of the farmers are women is a similar phenomenon. This is often found where certain crops of livestock are the traditional province of women farmers, and male agricultural extension officers do not communicate with them effectively. Until more women are selected and trained as agricultural extension staff, this problem is not likely to be overcome.

Another personnel problem relates to training. If a new young recruit into an agricultural extension organization has never lived or worked on a farm, as is most often the case, it takes a great deal of practical hands-on experience and practical training before that person is likely to appreciate what is really going on in agriculture. Even then, the trainee may lack commitment and enthusiasm for field extension work. Most training for such personnel tends to be too short, too theoretical, and too centered on lectures and books.

An outstanding world example of an effective training program is the International Rice Research Institute (IRRI). Trainers at IRRI have insisted that all trainees have field experience and actually transplant rice or plant seed or harvest grain with their own hands. Whether the program was one week or several months in duration, a practical 'hands-on' dimension has always been part of it. This in itself has been a powerful force throughout Asia in helping agricultural extension workers successfully spread the improved germ plasm developed by their research colleagues. Whether the impact of the training program or the new germ plasm was more significant would be hard to prove, but agricultural training at IRRI should be an example to teachers everywhere who are training new agricultural extension staff.

Finally, with respect to personnel, reward systems have been a key problem. Vigorous, dedicated, competent young men and women are not likely to stay in remote rural locations doing extension work unless there are more rewards and fewer punishments for such work. The family will not let them stay and the pressure to move on to a different career, or into the headquarters city, is usually too great to withstand.

On the program side, there are also constraints. If the same program is offered for the whole country, it is not likely to fit well in many different locations. If the program is controlled locally, it is more likely to be relevant, but may not please political leaders in the center. This is related not only to relevance, but to implementability. A great extension target which is achieved, but not relevant to the local situation, must be considered a failure. And any extension program which is not implementable is not going to be a success.

Beyond the techniques which enable extension education – the choice of appropriate communication channels, attractive treatments for relevant messages, and timing to enhance impact – the need for an internal discipline among agricultural extension field personnel, their supervisors, and technical specialists who provide substance to their programs is paramount. This discipline can be seen in the vigor with which staff approach their work. It can be seen in the extent to which they are physically present on farms, listening to farm men and women. It can be seen in the hours they work. It can be seen in the rewards received ... or not received ... by those who stay in the field and work where the farmers are, compared with those who gravitate to the central administration. That discipline - more than any other criterion - is a factor for success in agricultural extension education.

TYPES OF SYSTEMS

Although there are many different types of agricultural extension education systems, I find it useful to divide them into two basic categories [9]. One category may be called the agricultural extension delivery system and the other category is termed the agricultural extension acquisition system. Here in the USA, most agricultural extension work started as acquisition systems (i.e. the early County Farm Bureau). Only later did delivery systems emerge.

The main idea of a delivery system type of agricultural extension organization is that there is a body of information which farmers need. The organization either has this information or can get it, and the purpose of the organization is to deliver the information to farmers. Besides information, there may be other inputs such as fertilizer, seed, or credit. Government Agricultural Extension in India, Pakistan, and Bangladesh, along with Thailand and others, are examples of delivery systems. They are almost always a part of a Ministry of Agriculture. Staff at all levels are government officers. Program targets, goals, and objectives tend to be fixed by the governments and strategies, tactics, and other aspects of implementation are decided centrally.

Acquisition systems in agricultural extension are very different. Here the main idea is that groups of farmers, organized one way or another, can reach out beyond their villages, and acquire the information they need. These are usually smaller organizations, like Farmers Associations, Small Farmers Groups, Brigades, or Village-Level Cooperatives. In some countries, like Malaysia and Nepal, these are found operating along side of the Ministry of Agriculture Extension System. Indonesia has experimented with several different types, and at their best, the Communes and Brigades of the Peoples' Republic of China are acquisition systems.

Unlike delivery systems, staff of acquisition systems are employed by local organizations, sometimes enjoying cost-sharing arrangements with larger government units or other outside sponsors. Program targets, goals, and objectives are fixed by members of the group themselves and strategies, tactics, and other aspects of implementation are different from one village to the next, and from region to region within a country.

Among delivery systems, some deal with all aspects of agriculture (including livestock, fisheries, and forestry) while others are more specialized. The rubber extension group in Malaysia or the tobacco organization in Bangladesh are examples of single crop agricultural extension delivery systems. Others deal with multiple crops and livestock but tend to take them one commodity at a time. There is now an increasing number of agricultural extension efforts dealing with the farming system as a whole but these tend to be small, experimental, and widely scattered, sometimes combining research and extension activities. There is also a considerable amount of agricultural extension 'delivered' by even more generalized rural development agencies, often in integrated rural development projects, and usually associated with a ministry of local development or community affairs.

Acquisition systems are generally concerned with a broader range of agricultural subjects, shifting their focus from time to time as village problems change or as new needs arise. The World Conference on Agrarian Reform and Rural Development (WCARRD), organized by FAO in Rome in 1979, suggested the clear advantage of acquisition systems. The analysis of country experience in the implementation of the WCARRD Program of Action, just published by FAO, concludes that 'an increasing number of countries are showing interest and are trying the participatory approach in extension as a way of reaching large numbers of small farmers more effectively ...' That report also points out that 'an increasing number of countries are specifying small farmers and, in a few cases, including women as the main target clientele of the extension activities' [11].

Current patterns of agricultural extension also include variations in who controls the agricultural extension system – such as governments, commodity groups, or rural people (organized in various ways) – and variations in intended beneficiaries of agricultural extension – such as consumers, producers, industry, or even broader national interests. There are also different patterns of the level of discipline, or the index of seriousness. These separate one agricultural extension system from another, or sometimes just one section of a national agricultural extension system from other sections.

But rather than continue with the differences in patterns, the remainder of this discussion will focus on some of the most relevant and most illuminating issues concerning effectiveness in diffusion of new agricultural technology and in applicability to small farms in the so-called 'developing' countries.

ISSUES

A major issue which is beginning to emerge is that of whose interests are served by the agricultural extension system. The main conflict is between the urban consumer interest and the rural producer interest. Most agricultural extension systems are dominated by the political policy of 'cheap food in the cities'. The alternative interest of improved quality of life for rural people is always part of the rhetoric, but not as well evidenced by programs, personnel, or the operating doctrine of the extension system. It is in the urban consumer interest for farmers to produce more – to produce a surplus. It is in the rural farm family interest to consume more.

Problems arise for agricultural extension when central targets are set with a production orientation without price policies to match. Over the years, Japan has overcome these problems by keeping the domestic price of rice to farmers high enough to encourage production. Production oriented agricultural extension programs have a much greater chance of success when those who produce more are sufficiently rewarded that they may also consume more.

Another example of the issue of whose interests are served relates to differences between large commercial farms, on the one hand, and small self-sufficient farms on the other. In spite of aspirations of some scientists involved in agricultural research, most technologies are not scale neutral. And most productivity-enhancing technology developed by the international agricultural research community assumes the type of farming system which purchases its inputs from off the farm and sells its outputs to commercial marketing channels. However, most of the small farming systems, particularly in Asia, produce their own inputs, and consume their own outputs. They are not completely self-sufficient but they are highly self-sufficient. Further, the large farms tend to specialize in one or only a few crops, or classes of livestock. Small farms tend to have mixed crop and livestock systems, with many different cereal grains, farm animals, fruits and vegetables. With technology generated to meet the needs and solve the problems of large-scale commercial agriculture, it has been very difficult for agricultural extension systems to have much impact on small mixed self-sufficient farming systems.

A second issue revolves around unrealistic expectations for agricultural extension. Agricultural extension is often blamed when unrealistic expectations result in farmers' refusal to adopt new technologies. One example has already been presented with respect to price policy. It is unrealistic to expect agricultural extension to be able to convince farmers to add mineral fertilizer, for example, when the price of fertilizer, compared to the price of the crop, is such that the farmers who follow the recommendation would lose by doing so. If a technological innovation is not supported by appropriate price policy, agricultural extension cannot convince farmers to accept the innovation. Another example appears when the new technology simply does not fit the type of farming system or the agro-ecological-economic environment. Attempts to introduce artificial insemination in dairy cattle face this problem where communication and transportation infrastructure do not facilitate this technology.

An additional unrealistic expectation is that extension staff can somehow bring about an increase in production without appropriate technology being generated or made available to them. Where they have nothing practical and useful to offer to farm families, their work is not likely to result in increased production.

A final example is found where agricultural research systems respond to rewards from outside the country - sometimes rewards from the international agricultural research system instead of the needs of rural people in its own country. If the research system attempts to pressure agricultural extension to promote its 'findings', extension is likely to fail. This has happened when short stemmed cereal grain varieties were introduced in regions with very high populations of ruminant animals. Since the farmer absolutely needs the straw as fodder for the animals, the extension effort failed.

A third vital issue is that of <u>control</u> of agricultural extension. Earlier this manuscript quoted Thoreau's statement that there are thousands hacking away at the branches of evil for every one who is cutting at the roots. From the perspective of this author, the branches which are often analyzed in agricultural extension include: weak impact on farmers, lack of discipline, lack of esprit de corps, lack of recognition and reputation. The roots, however, are control of personnel and program.

Who controls agricultural extension can be determined by asking local agricultural extension officers who pays their salary. Also, if they receive an increase in their salary, who decides how much it should be and how often they should receive it? If their answer is the central government, or the national rubber board (for example), then they are part of a delivery system. If their answer is that the members of their cooperative, or the farm families of their district decide these matters, then it is an acquisition system. If the answer is somewhere in between, then control of the organization is somewhere in between.

Where agricultural extension education is not controlled by the farm families it is expected to serve, it is likely to have an inappropriate program with targets which do not fit the situation and implementation means which fail. And where agricultural extension is not controlled by its clientele, the personnel - that is the extension staff - are likely to be poorly paid, poorly trained, poorly managed - and not very effective.

A short-term solution to this symptom has been the Extension Training and Visit system. It does bring discipline to the system and, under certain conditions, has increased effectiveness. But it fades without appropriate control, as neither program nor staff are likely to maintain effectiveness if control is centralized. It becomes plagued by the conflicting agendas of professional agriculturalists and of farmers - and with control in the hands of the professionals, neither program nor personnel tend to be responsive to the needs and interests of farmers.

The long-term solution, in the view of this author, is to place control in the hands of the so-called target groups - the farming families themselves. When they organize their own acquisition systems, the program is likely to fit their needs and interests. When national groupings of farmers' acquisition systems are organized to reflect these needs and interests, they can exert influence on the agenda of agricultural research, as well as on national policy.

A fourth issue, found throughout all levels of agricultural education, may be labeled the issue of practical experience versus merely literacy. Too often agriculturalists have little practical training or experience. They have memorized the text books, and offer only a 'literacy' understanding of farming. At the local level they are little help to farmers. At regional levels, if they are training agricultural extension officers, they are just as unsuccessful.

The final issue is the issue of women in agriculture and food systems. Visibly evident throughout Asia, but also true in other parts of the world, this is a serious anomaly for agricultural extension, while most of the farmers are women, and most of the extension personnel are men.

Some facts from Nepal are illustrative of similar phenomena in many other countries. There, rural women's total work burden is extremely high, at an average of 10.81 hours per day, compared to 7.51 hours per day for men. Rural Nepalese women not only contribute more time, but also generate more income than men for the total household economy. Women are primarily responsible for the farm enterprise both in terms of labor contribution (9.9 hours versus 5.86 hours per day for men) and management decisions. The evidence for this, documented in the last decade, is compelling [12]. We witness a high proportion of effort with the husbands of farmers, and little direct effort either with women farmers or with whole farming families.

There are gender-sensitive issues in the selection of future extension personnel, in the training of agriculturalists, and in the agenda of agricultural research. The opportunity for significant improvements in the effectiveness of agricultural extension education may hinge on the ability of the systems to respond appropriately to the challenging reality of women in agriculture.

Personally, I do not believe that separate projects designed for rural women are going to solve the problem. What is more desirable is a dimension of focus and concern for women and families in every aspect of agricultural extension education. With this focus, whether an extension system is working on rice production or keeping dairy or aquaculture, or any other farming topic, it could be more effective. A much higher proportion of field agriculturalists would be women. They could contact farming women directly.

This issue is not going to be resolved quickly. It is a long range matter, involving changes in attitudes of people. This calls for changes in admissions to the schools and colleges of agriculture, as well as changes in recruitment practices of agricultural extension organizations. But changes in this area offer great promise to increased relevance and increased effectiveness in agricultural extension organizations in most of the world.

In conclusion, it may be beneficial to conclude this discussion of agricultural extension by turning to some advice given by a Bengali philosopher who was himself an agricultural extensionist and a teacher of rural development. It was Rabindra Nath Tagore who wrote:

A lamp cannot light another lamp
  unless it itself is also lit.
A teacher cannot truly teach
  unless he himself is also learning.

REFERENCES

Acharya, M. and Bennett, L. (1981) The Rural Women of Nepal, An Aggregate Analysis and Summary of 8 Village Studies, Volume II, part 9, Tribhuvan University, Kathmandu, Nepal: Center for Economic Development and Administration.

Axinn, G.H. (1977) 'Agricultural Research, Extension Services, and Field Station, in The International Encyclopedia of Higher Education, San Francisco and London: Jossey-Bass, pp. 421-54.

Axinn, G.H. (1978) New Strategies for Rural Development, Kathmandu, Nepal: Rural Life Associates.

Axinn, N.W. (1981) Mediating Conflicting Agendas in the Development Process. Proceedings of the Conference, 'Responding to the Needs of Rural Women, South-East Consortium for International Development', Frankfort: Kentucky State University.

Axinn, G.H. and Axinn, N.W. (1976) Non-Formal Education and Rural Development, Program of Studies in Non-Formal Education, Supplementary Paper No. 7, East Lansing, MI: Michigan State University.

Axinn, N.W. and Axinn, G.H. 'Energy and Food Relationships in Developing Countries', Chapter 6 in Food and Energy Relationships, C.W. Hall and D. Pimental (eds), Academic Press, Inc., pp. 121-46.

Axinn, G.H. and Thorat, S.S. (1972) Modernizing World Agriculture: A Comparative Study of Extension Systems, N.Y., New Delhi, Oxford and IBH: Praeger Publishers.

Chambers, R. (1980) 'Rural Poverty Unperceived: Problems and Remedies', World Bank Staff Working Paper, No. 400, Washington, DC: The World Bank.

El Chonemy, M.R. (ed.) (1984) How Development Strategies Benefit the Rural Poor (WCARRD Follow-up Program). Rome: Food and Agriculture Organization of the United Nations.

French, J.H. (1982) Introduction to the Small Farmers Development Project, Bangkok, Thailand: DTCP for FAO.

Roling, N. (1984) 'Appropriate Opportunities as well as Appropriate Technology', CERES, FAO Review on Agriculture and Development, No. 97 (Vol. 17, No. 1) Jan.-Feb., pp. 15-19.

Swanson, B.E. and Rassi, J. (1981) International Directory on National Extension Systems, Urbana-Champaign, IL: University of Illinois.

Chapter Eight

THE IARCs AND THEIR IMPACT ON NATIONAL
RESEARCH AND EXTENSION PROGRAMS

Robert E. Evenson
Yale University

The first International Agricultural Research Center (IARC),
IRRI, is now 25 years old [1]. Several other IARCs have
been in place for more than 15 years. A number of important
changes have taken place, both in the development of the
IARCs and in the building of national research and extension
capacity in the developing world over this period [2]. This
paper reports the findings of a study that seeks to determine
whether the development of the IARC system has produced a
measurable impact on the size and character of national agricultural research and extension programs.

The first section of the paper provides a descriptive
summary of national research and extension spending since
1959. The second section discusses the rationale for national
research and extension investment. The third section summarizes calculations based on an econometric study of the
determinants of investment in national research and extension
programs and draws inferences regarding IARC impact.

I.   A DESCRIPTIVE SUMMARY OF NATIONAL AND
     INTERNATIONAL PROGRAM DEVELOPMENT

National investment in agricultural research and extension
programs has grown at an impressive rate in the past 25
years [3]. Tables 8.1 and 8.2 summarize this investment. It
may be seen that, in 1980 constant dollars, research spending
in developing countries increased from 1959 to 1980 by a
multiple of 5.8 in Latin America, 6.9 in Asia, and 3.6 in
Africa. The comparable spending multiples for extension
investment were 6.4 for Latin America, 3.5 for Asia, and 2.2
for Africa. Scientist man-year (SMY) multiples were lower
than spending multiples (6.0 for Latin America, 4.1 for Asia,
4.2 for Africa) reflecting rising real costs per SMY. (For
extension workers the multiples were 6.8 for Latin America,
1.8 for Asia, 2.9 for Africa.)

Table 8.1: International agricultural research expenditures and scientific manpower 1959, 1970 and 1980

Region/Subregion	EXPENDITURES (000 Constant 1980 US$)			MANPOWER (Scientist Man-years)		
	1959	1970	1980	1959	1970	1980
Western Europe	274,984	918,634	1,489,588	6,251	12,547	19,540
Northern Europe	94,718	230,135	409,527	1,818	4,409	8,027
Central Europe	141,054	563,334	871,233	2,888	5,721	8,827
Southern Europe	39,212	125,165	203,828	1,545	2,417	2,886
Eastern Europe and USSR	568,284	1,282,212	1,492,783	17,701	43,709	51,614
Eastern Europe	195,896	436,094	553,400	5,701	16,009	20,220
USSR	372,388	846,118	939,383	12,000	27,700	31,394
North America and Oceania	760,466	1,485,043	1,722,390	8,449	11,683	13,607
North America	668,889	1,221,006	1,335,584	6,690	8,575	10,305
Oceania	91,577	264,037	386,806	1,759	3,113	3,302
Latin America	79,556	216,018	462,631	1,425	4,880	8,534
Temperate South America	31,088	57,119	80,247	364	1,022	1,527
Tropical South America	34,792	128,958	269,443	570	2,698	4,840
Caribbean and Central America	13,676	29,941	112,941	491	1,160	2,167

Africa	119,149	251,572	424,757	1,919	3,849	8,088
North Africa	20,789	49,703	62,037	590	1,122	2,340
West Africa	44,333	91,899	205,737	412	952	2,466
East Africa	12,740	49,218	75,156	221	684	1,632
Southern Africa	41,287	60,752	81,827	696	1,091	1,650
Asia	261,114	1,205,116	1,797,894	11,418	31,837	46,656
West Asia	24,427	70,676	125,465	457	1,606	2,329
South Asia	32,024	72,573	190,931	1,433	2,569	5,690
Southeast Asia	9,028	37,405	103,249	441	1,692	4,102
East Asia	141,469	521,971	734,694	7,837	13,720	17,262
China	54,166	502,491	643,555	1,250	12,250	17,273
World Total	2,063,553	5,358,595	7,390,043	47,163	108,510	148,039

Source: Boyce, J.K. and R.E. Evenson, National and International Agricultural Research and Extension Programs (New York: The Agricultural Development Council, 1975), and M. Ann Judd, James K. Boyce, and Robert E. Evenson, 'Investing in Agricultural Supply' (Discussion Paper No. 442, Yale University, Economic Growth Center, 1983).

Table 8.2: Agricultural extension expenditures and manpower

Region/Subregion	EXPENDITURES (000 Constant 1980 US$)			MANPOWER (Workers)		
	1959	1970	1980	1959	1970	1980
Western Europe	234,016	457,675	514,305	15,988	24,388	27,881
Northern Europe	112,983	187,144	201,366	4,793	5,638	6,241
Central Europe	103,082	199,191	236,834	7,865	13,046	14,421
Southern Europe	17,950	71,340	76,105	3,330	5,704	7,219
Eastern Europe and USSR	367,329	562,935	750,301	29,000	43,000	55,000
Eastern Europe	126,624	191,460	278,149	9,340	15,749	21,546
USSR	240,705	371,475	472,152	19,660	27,251	33,454
North America and Oceania	383,358	601,950	760,155	13,580	15,113	14,966
North America	332,892	511,883	634,201	11,500	12,550	12,235
Oceania	50,466	90,067	125,954	2,080	2,563	2,731
Latin America	61,451	205,971	396,944	3,353	10,782	22,835
Temperate South America	5,741	44,242	44,379	205	1,056	1,292
Tropical South America	47,296	136,943	294,654	2,369	7,591	16,038
Caribbean and Central America	8,414	24,786	57,911	779	2,135	5,505

Africa	237,883	481,096	514,671	28,700	58,700	79,875
North Africa	84,634	176,498	172,910	7,500	14,750	22,453
West Africa	53,600	181,324	204,982	9,000	22,000	29,478
East Africa	39,496	86,096	106,030	9,000	18,750	24,211
Southern Africa	60,153	37,178	30,749	3,200	3,200	3,733
Asia	143,876	412,937	507,113	86,900	142,500	148,780
West Asia	28,211	97,315	119,780	7,000	18,800	16,535
South Asia	56,422	87,727	82,194	57,000	74,000	80,958
Southeast Asia	19,747	55,441	63,959	9,500	30,500	33,987
East Asia	39,496	172,454	241,180	13,400	19,200	17,300
China	n.a.	n.a.	n.a.	n.a.	n.a.	n.a.
World Total	1,427,913	2,722,564	3,443,489	177,521	294,483	349,337

Source: Boyce, J.K. and R.E. Evenson, National and International Agricultural Research and Extension Programs (New York: The Agricultural Development Council, 1975), and M. Ann Judd, James K. Boyce, and Robert E. Evenson, 'Investing in Agricultural Supply' (Discussion Paper No. 442, Yale University, Economic Growth Center, 1983).

Table 8.3 shows how research and extension 'spending intensities', i.e. spending as a percentage of the domestic value of agricultural product (G.D.P.) has changed from 1959 to 1980. These data show that in 1959 the low-income and middle-income developing countries were approximately twice as spending-intensive for extension as for research [4]. The reverse was true for the industrialized countries. The rapid growth in spending intensities for research from 1959 to 1980 combined with little or no growth in extension intensities in the 1970s produced roughly equal spending intensities for research and extension in most developing countries.

Table 8.4 provides comparable data for 'manpower intensities' (i.e. ratios of manpower to G.D.P.). For research the same general pattern reflected in spending intensities is reflected in the manpower intensities. Because spending per SMY is lower in developing countries they fare better by this measure. The difference between the low-income and industrialized countries is much reduced.

For extension, the picture is quite different. By 1959 low-income developing countries had attained very high extension manpower intensities; five to seven times greater than those attained in industrialized countries. By 1980, with a slight decline in these intensities for industrialized countries, the difference was even greater. Middle-income and semi-industrialized countries also increased their extension intensities.

These manpower intensities should not be interpreted as though there were no differences in the quality of manpower between countries. There is little doubt that the general levels of training of both scientists and extension workers vary between countries and are lower in the developing countries. However, the differences are not as great as is generally supposed. There is also little indication that these differences have changed as research and extension spending has increased. These data do not include 'extension type' spending associated with Rural Development Projects in developing countries. Were such data to be tabulated and included as extension spending, the magnitude of the differences in spending on extension relative to research in the developing countries would be even greater.

Table 8.5 provides further insight into the motivation for the high extension manpower intensities in developing countries. It shows expenditure/manpower ratios for research and extension. These ratios include salaries of scientists and extension workers and related costs, including laboratory costs and the costs of technicians. The ratio of research costs to extension costs is as much as 20 to 1 for the low-income developing countries and only 3 to 1 or so for the industrialized countries. Some of this difference is a quality difference (extension workers have quite advanced training in most industrialized countries and may have little training in low-income countries), and some is due to real cost differences. Many low-income countries do not have the capacity to train agricultural scientists and must incur high costs to train researchers and to purchase scientific equipment.

Table 8.6 reports data on spending by commodity in the form of spending intensities. With few exceptions, developing countries cannot provide a commodity breakdown for their research spending. They do well to provide data on total spending. It is possible, however, to obtain publications data from the CAB Abstract system by commodity orientation. This was done for each of 25 countries for the two periods 1972-5 and 1976-80. These data were then standardized into equal cost units utilizing Brazilian data. For Brazil real spending by commodity and CAB publications data were available. It was thus possible to standardize publications into cost equivalent units. Standardized publications were then used to allocate actual expenditures to commodities.

The date show that spending intensities differ greatly by commodity in the 25 country sample (these 25 countries account for approximately 90 per cent of total production in developing countries, excluding China). Spending intensities are low for coconuts, sweet potatoes and cassava and high for cocoa, coffee and livestock. The table also shows that the IARCs account for relatively low shares of the total research on the commodities they work on. Since expenditures per SMY are very high in the IARCs (about 4-6 times the average for national spending), the IARCs are much less significant in terms of their share of scientific manpower devoted to these commodities.

II. SPECIFYING THE DETERMINANTS OF INVESTMENT IN RESEARCH AND EXTENSION

If IARC impacts on national research and extension spending are to be measured a specification relating national spending to 'determinants', including IARC investment, is required. Such a specification should be consistent with economic logic and political reality. Since IARC investments are commodity based, it is natural to develop the specification for spending by commodity.

The specification developed here is motivated by a project evaluation or planning perspective modified by political constraints. The specification includes variables that a rational planner would use to guide optimal investment. It also includes variables that reflect the political power of interest groups and political constraints.

The independent variables in the analysis are the variables measuring national research spending and national extension spending.

The model by which this spending is determined is constructed in stages. The first stage is motivated by sup-

Table 8.3: Research and extension expenditures as a percentage of the value of agricultural product

Subregion	Public Sector Agricultural Research Expenditures			Public Sector Agricultural Extension Expenditures		
	1959	1970	1980	1959	1970	1980
Northern Europe	.55	1.05	1.60	.65	.85	.84
Central Europe	.39	1.20	1.54	.29	.42	.45
Southern Europe	.24	.61	.74	.11	.35	.28
Eastern Europe	.50	.81	.78	.32	.36	.40
USSR	.43	.73	.70	.28	.32	.35
Oceania	.99	2.24	2.83	.42	.76	.98
North America	.84	1.27	1.09	.42	.53	.56
Temperate South America	.39	.64	.70	.07	.50	.43
Tropical South America	.25	.67	.98	.34	.71	1.19
Caribbean & Central America	.15	.22	.63	.09	.18	.33
North Africa	.31	.62	.59	1.27	2.21	1.71
West Africa	.37	.61	1.19	.58	1.24	1.28
East Africa	.19	.53	.81	.67	.88	1.16
Southern Africa	1.13	1.10	1.23	1.64	.67	.46
West Asia	.18	.37	.47	.25	.57	.51
South Asia	.12	.19	.43	.20	.23	.20
Southeast Asia	.10	.28	.52	.24	.37	.36

East Asia	.69	2.01	2.44	.19	.67	.85
China	.09	.68	.56	n.a.	n.a.	n.a.
Country Group*						
Low-income Developing	.15	.27	.50	.30	.43	.44
Middle-income Developing	.29	.57	.81	.60	1.01	.92
Semi-industrialized	.29	.54	.73	.29	.51	.59
Industrialized	.68	1.37	1.50	.38	.57	.62
Planned	.33	.73	.66	–	–	–
Planned – excluding China	.45	.75	.73	.29	.33	.36

* For definition of Country Groups see Note 2

Source: Tables 8.1 and 8.2; and USDA, Indices of Agricultural Production, various issues.

Table 8.4: Research and extension manpower relative to the value of agricultural product

Subregion	SMYs per 10 million (Constant 1980) Dollars Agricultural Product			Extension Workers per 10 million (Constant 1980) Dollars Agricultural Product		
	1959	1970	1980	1959	1970	1980
Northern Europe	1.05	2.01	3.14	2.76	2.56	2.61
Central Europe	.80	1.21	1.56	2.19	2.77	2.73
Southern Europe	.93	1.17	.96	2.00	2.76	2.69
Eastern Europe	1.44	2.97	2.84	2.36	2.88	3.13
USSR	1.38	2.37	2.34	2.26	2.33	2.50
Oceania	1.91	2.64	2.43	2.26	2.17	2.11
North America	.84	.89	.84	1.44	1.31	1.08
Temperate South America	.46	1.15	1.32	.26	1.19	1.26
Tropical South America	.41	1.41	1.77	1.71	3.95	6.46
Caribbean & Central America	.53	.86	1.20	.82	1.53	3.12
North Africa	.91	1.44	4.24	18.83	28.45	22.23
West Africa	.33	.61	1.42	7.61	14.01	18.08
East Africa	.32	.77	1.76	16.28	22.41	26.64
Southern Africa	1.90	1.96	2.47	8.73	5.94	5.62

West Asia	.33	.84	.88	4.39	7.25	6.54
South Asia	.50	.65	1.29	20.83	19.51	19.53
Southeast Asia	.47	1.28	2.07	9.81	13.07	19.72
East Asia	3.80	5.29	5.72	6.57	7.05	6.13
China	.22	1.66	1.49	n.a.	n.a.	n.a.
Country Group						
Low-income Developing	.43	.67	1.40	18.14	18.61	20.43
Middle-income Developing	.69	1.31	2.40	8.89	14.68	15.98
Semi-industrialized	.70	1.21	1.36	2.80	4.95	5.21
Industrialized	1.24	1.71	1.85	2.37	2.31	2.12
Planned	1.02	2.27	2.13	–	–	–
Planned – excluding China	1.40	2.54	2.50	2.29	2.49	2.63

Source: See Tables 8.1, 8.2 and 8.3.

Table 8.5: Expenditures per SMY/extension worker

Region/Subregion	Research Expenditures per SMY (000 Constant 1980 US$)			Extension Expenditures per Extension Worker (000 Constant 1980 US$)		
	1959	1970	1980	1959	1970	1980
Western Europe	44	73	76	15	19	18
Northern Europe	52	52	51	24	33	32
Central Europe	49	98	99	13	15	16
Southern Europe	25	52	78	5	13	11
Eastern Europe and USSR	32	29	29	13	13	14
Eastern Europe	34	27	27	14	12	13
USSR	31	31	30	12	14	14
North America and Oceania	90	127	127	28	40	51
North America	100	142	130	29	41	52
Oceania	52	85	117	24	35	46
Latin America	56	44	54	18	19	18
Temperate South America	85	56	53	28	42	34
Tropical South America	61	48	56	20	18	18
Caribbean and Central America	28	26	52	11	12	11

Africa	62	65	53	8	8	6
North Africa	35	44	27	11	12	8
West Africa	108	97	83	6	8	7
East Africa	58	72	46	4	5	4
Southern Africa	59	56	50	19	12	8
Asia	23	38	39	2	3	3
West Asia	53	44	54	4	5	7
South Asia	22	28	34	1	1	1
Southeast Asia	20	22	25	2	2	2
East Asia	18	38	43	3	9	14
China	43	41	37	n.a.	n.a.	n.a.
Country Group						
Low-income Developing	34	40	47	2	2	2
Middle-income Developing	42	44	47	7	7	6
Semi-industrialized	41	45	46	10	10	11
Industrialized	55	80	93	16	25	29
Planned	33	32	31	-	-	-
Planned excluding China	31	25	30	13	13	14

Source: See Tables 8.1 and 8.2.

Table 8.6: Research as a percentage of the value of product, by commodity, average 1972-9 period, 25 countries

Commodity	REGION			All Countries	Spending by International Centers	Ratio IARC Spending to Total
	Africa	Asia	Latin America			
Wheat	1.30	.32	1.04	.51	.02	.04
Rice	1.05	.21	.41	.25	.02	.07
Maize	.44	.21	.18	.23	.03	.11
Cotton	.23	.17	.23	.21	-	-
Sugar	1.06	.13	.48	.27	-	-
Soybeans	23.59*	2.33	.68	1.06	-	-
Cassava	.09	.06	.19	.11	.02	.15
Field Beans	1.65	.08	.60	.32	.04	.11
Citrus	.88	.51	.57	.52	-	-
Cocoa	2.75	14.17*	1.57	1.69	-	-
Potatoes	.21	.19	.43	.29	.08	.21
Sweet Potatoes	.06	.08	.19	.07	-	-
Vegetables	1 (=1.56)	.41	1.13	.73	-	-
Bananas	.27	.20	.64	.27	-	-
Coffee	3.12	1.25	.92	1.18	-	-
Groundnut	.57	.12	.60	.25	.005	.02
Coconut	.07	.03	.10	.04	-	-
Beef	1.82	.65	.67	1.36	.02	.02
Pork	2.56	.39	.60	1.25	.02	.02
Poultry	1.99	.32	1.12	1.64	-	-
Other Livestock	1.81	.89	.42	.71	-	-

Sources: M. Ann Judd, James K. Boyce, and Robert E. Evenson, 'Investing in Agricultural Supply' (Discussion Paper No. 442, Yale University, Economic Growth Center, 1983); and USDA, Indices of Agricultural Production, various issues. * Ratios are high because production is very low.

posing that a planner is attempting to maximize the economic surplus, (i.e. both consumers' and producers' surplus) associated with the research or extension program. In the second stage the planner takes international transfer conditions into account. In the third, the planner takes political constraints into account.

Now consider the first stage of the planner's problem. A given research program can be expected to lower production costs per unit of production. The more units over which cost can be lowered, the higher the optimal level of research. Each commodity and each geo-climate region present different research problems to some degree. Hence units of production should be measured on a commodity-region basis. National research spending is expected to rise as both production and diversity increase.

For some (perhaps most) research programs a 'minimum critical mass' of research effort may be required for an effective program. If so there will be a threshold level of production below which a research program cannot be justified. Small diverse countries are more likely than larger countries to face these problems.

Planners will respond to price variables reflecting prices of alternative sources of growth in supply. The ratio of expenditures per SMY to expenditures per extension worker is designed to reflect the relative costs of pursuing growth through extension investment. It is expected that when the price of research resources falls relative to extension resources, more spending in research will take place. The ratio of the arable land currently to arable land six years previously is designed to reflect the price of supply growth via land expansion. When the change in arable land is small, reflecting land exhaustion, more spending on research is expected.

Now turn to the second stage of the problem. The planner recognizes that technology may 'spill-in' from other countries and from IARCs. He also recognizes, however, that the potential spill-in technology was designed for on 'targeted' to geo-climate conditions in other countries. Other national programs will be targeting their research programs to their own geo-climate conditions. The IARCs may target to a broader range of conditions than are extant in their host countries, but in practice they lack the resources to provide technology targeted to more than a limited range of environments. Thus, the planner will find that some technology available on the international market is directly suited to use (i.e. it is targeted to domestic conditions) but that much new technology (and related research findings) is 'mismatched', i.e. it is targeted to geo-climate conditions differing from those of the country. The planner's response to closely matched technology from abroad will be to reduce domestic research investment since domestic research is a substitute for matched technology from abroad (extension spending may be inversed). The planner's response to mismatched technology from abroad will be to increase domestic research investment since this mismatched technology offers domestic researchers an opportunity for modification and adaptation of the mismatched technology to domestic conditions. Of course, if the mismatch is too great it will not offer such opportunities.

We would then expect planners to exhibit a mixed response to technology from abroad. On the one hand, they will 'free ride' on the research of IARCs and neighboring countries to the extent that they see these research units as producing closely matched technology with little scope for adaptation. On the other hand, they will respond with increased adaptive research to the extent that they see these units producing mismatched technology offering adaptation opportunities and to the extent that these units are producing 'pre-technology' scientific discoveries that also enhance the productiveness of their own systems.

Finally, the planner will respond to political constraints. Most countries implicitly place a higher value on international exchange than on domestic production. A unit of product that saves or earns foreign exchange is valued more highly than one that does not. A planner will respond to this by investing more in research on commodities that save or earn foreign exchange. Many countries intervene in agricultural markets. The ratio of prices paid for urea fertilizer to prices raised for rice is a measure of this intervention. A planner might attempt to 'compensate' for some types of intervention by spending more or less on research.

Planners will also respond to pressure from interest groups. They may, for example, respond to urban pressure groups by shifting resources from research to competing investments even though urban consumers are the major beneficiaries of agricultural research [5]. High proportions of the labor force in agriculture are usually associated with weak political power of rural people. If so, this could reduce spending on research and extension.

III. POLICY IMPLICATIONS OF INVESTMENT ESTIMATES

The results of an econometric exercise based on the model discussed above are reported in detail in Evenson, 1986. They have substantial policy relevance. They show a considerable degree of consistency with rational planning on the part of national governments. However, another large body of evidence (see Evenson, Waggoner and Ruttan, 1979 and Ruttan, 1983) shows that research investments have produced extraordinarily high returns in terms of the increased agricultural output associated with research programs. The implication is that there is general underinvestment in research.

With this in mind, it is useful to examine the marginal impacts of alternative policy-related activities on national research and extension spending. Table 8.7 reports a number of such calculations based on the regression estimates reported in Evenson (1986).

Table 8.7 shows that as commodity production increases, spending on both research and extension in increased. In percentage terms these estimates show that if production is 10 per cent higher, spending on both research and extension goes up by from 5.5 to 6 per cent. The fact that it does not increase proportionately with production indicates that a type of 'scale economy' is perceived by these governments in their spending decisions. This is partially related to a minimum or threshold research or extension effort. This works to the disadvantage of the relatively small country.

The table also shows that when the commodity being produced is exported research spending per unit of product is 1.39 times as high for field crops and 1.54 times as high as for livestock and horticultural crops as it is for non-traded commodities. When the commodity is imported, spending per unit of product is 1.29 times as high for field crops and over 4 times as high for livestock and horticultural crops (where imports are generally low). The policy implication for these calculations is not that traded commodities receive too much research attention but that non-traded commodities almost certainly receive too little attention.

The results show that the research of geo-climate neighbors stimulate research spending for all field crops (except wheat, potatoes and sweet potatoes) but that it actually causes a small decline in extension spending. This indicates that countries are not seeking to 'free ride' on the technology produced by other countries. Instead they respond to research opportunities by spending more on their own research programs.

The estimates also show that the relative costs of research and extension do matter. A 10% fall in the costs of research per scientist man-year causes spending on field crops research to rise slightly. This really means that a country will increase the number of scientist man-years by slightly less than 10% - a very appreciable rise. The response is lower in the case of livestock and field crops research. A 10% rise in the costs of extension workers leads to a 15% decrease in the numbers of extension workers employed.

The final calculations regarding aid and IARC spending are of most interest. The form of the model measuring IARC impacts was that the stock (i.e. cumulated expenditures in 1980 dollars) of IARC investment impacted on the annual flow of national research spending. Thus, a million dollar increment to IARC spending in 1978 would raise the value of

131

Table 8.7: Calculated impacts on national research and extension investment. Calculated impacts on spending in millions of 1980 dollars

Calculated from Table 10, Evenson 1985

| Policy Variable | Impacts on Research Spending | | Impacts on Extension Spending |
	Field Crops	Livestock & Horticulture Crops	
1 million $ added to commodity production	.00164	.00396	.00624
1 million $ added to commodity exports	.000634	.002277	.00695
1 million $ added to commodity imports	.000472	.01253	-.000937
1 added SMY by geo-climate neighbor	.0305	.01901	-.1792
Ten per cent decline in research costs per SMY or a ten per cent increase in extension costs per extension worker	-.0627	-.532	1.665
1 million dollars added to IARC research stock			
a) first year	.229	1.084	.105
b) after 10 years	2.290	10.840	1.050
1 million dollars general aid research	1.194	-.858	+.047
World Bank aid (to research or extension)	.285	-.063	1.468

Research Spending by Commodity (from Table 11, Evenson 1985)

	Maize	Sorghum	Millets	Rice	Wheat	Beans	Cassava	Ground-nuts	Potatoes	Sweet Potatoes
1 added SMY by geo-climate neighbor	.0217	.0307	.0355	.0121	-.0506	.0434	.0672	.0358	-.0069	-.0637
1 million dollars added to IARC investment										
a) first year	.225	.550	1.000	.168	1.725	.162	-.000	.650	1.050	.475
b) after 10 years	2.250	5.500	10.00	1.680	17.250	1.620	-.000	6.500	10.500	4.750

the IARC spending variable in 1978, 1979, etc. If this IARC spending was in the field crops it would stimulate $229,000 added national research investment in the first year (1978). (This is calculated as the total of the spending impacts in the 24 countries in the sample. Presumably the scope of influence is wider than for these 24 countries, so this is an underestimate of the effect.) By 1988 a total of $2,290,000 added national research investment would have been stimulated by the one million dollar expenditure in 1978. With the data at hand it is not really possible to estimate the deterioration of this effect. It is conservative to suppose that it will last only ten years (about the average time period for IARC investment in the data set).

The results for individual field crops also show investment impacts that are generally large. IARC investments of one million dollars in potatoes, sweet potatoes, wheat, sorghum and millets appear to stimulate an added million dollars in national spending within one or two years. Even for maize and rice the added national investment is significant.

This may be compared with the estimates for direct aid. They show that one million dollars in general aid increases field crop research by more than one million dollars but at the cost of reduced spending on livestock and field crop research. Thus, taking this displacement into account, only $336,000 net incremental research spending takes place for the one million dollar grant or loan. The same calculation made for World Bank aid shows an even more severe displacement effect. A million dollars in World Bank aid results in only a net increment to spending of $222,000. In rather sharp contrast, it appears the World Bank extension aid has a large stimulus effect on extension spending.

The aid inputs, it must be noted, are difficult to estimate and this will lead some policy makers to discount them. Most aid donors, however, are predisposed to believe that their aid has sufficient 'strings' and that it will not be displaced. Yet, most of it, in fact, is displaced and generally displacement is probably efficient. When accompanied by strong policy advice and pressure as in the case of World Bank extension aid (the T&V system) aid can have a large effect.

It appears, then, that the IARC system has had a significant and positive impact on national research (and extension) programs in the developing world. It has stimulated more spending in national systems and this impact is sufficiently large that an aid donor interested in stimulating national research spending actually received more stimulus from a grant to the IARC system than from a direct grant to a national system. The IARC system has probably also had a significant impact on more qualitative aspects of national research systems as well.

NOTES

1. See Oram and Blindish, 1981 for a detailed discussion of expenditures in the international system.
2. The development of national research and extension systems is documented in Judd, Boyce and Evenson, 1984 and Kislev and Evenson, 1975.
3. Judd, Boyce and Evenson, 1984 provide details. The tables in this chapter provide country tables summarizing changes in national system development.
4. The definition of country groups is that used by the Third World Bank in its World Development Report 1984.
5. Many studies show that while consumers are the major gainers from agricultural research, they are not strong supporters of research (see Evenson, 1982, and Rose-Ackerman and Evenson, 1983).

REFERENCES

Boyce, J.K. and Evenson, R.E. (1975) 'National and International Agricultural Research and Extension Programs', Agricultural Development Council.
Evenson, R.E. (1982) 'Observations on Brazilian Agricultural Research and Productivity', Revista de Economia Rural, 20:3, July/September.
Evenson, R.E. (1986) 'The IARCs: "Evidence of Impact on National Research and Extension Programs"' in G. Edward Schuh (ed.) Technology, Human Capital and the World Food Problems, University of Minnesota Agricultural Experiment Station, Misc. publication 37.
Evenson, R.E., Waggoner, P. and Ruttan, V.W. (1979) 'Economic Benefits from Research: An Example from Agriculture', Science, 205, September.
Guttman, J. 'Interest Groups and the Demand for Agricultural Research', Journal of Political Economy, 86, June.
Jamison, D.T. and Lau, J.J. (1982) Farmer Education and Farm Efficiency, Johns Hopkins University Press, Baltimore.
Judd, M.A., Boyce, J.K. and Evenson, R.E. (1984) 'Investing in Agricultural Supply', Economic Development and Cultural Change, forthcoming, E.G.C. Discussion Paper 442, Yale University, Economic Growth Center.
Kislev, Y. and Evenson, R. (1975) 'Investment in Agricultural Research and Extension, An International Survey', Economic Development and Cultural Change, 3, April.
Oram, P. and Bindlish, V. (1981) Resource Allocations to Agricultural Research: Trends in the 1970s (A Review of Third World Systems), International Food Policy Research Institute, Washington, DC.

Otsuka, K. (1978) 'Public Research and Price Distortion: Rice Sector in Japan', Ph.D. Dissertation, University of Chicago, 1979.

Rose-Ackerman, S. and Evenson, R.E. (1983) Public Support for Agricultural Research and Extension: A Political-Economic Analysis, Economic Growth Center, Yale University, New Haven, CT.

Ruttan, V.W. (1983) Agricultural Research Policy, World Bank World Development Report, Washington, DC.

Chapter Nine

TRAINING AND VISIT EXTENSION: BACK TO BASICS*

Daniel Benor
The World Bank

The Training and Visit (T&V) system of agricultural extension
has been widely adopted over the last decade and is now
being used in at least 40 countries. Many governments have
contributed significant resources to implementing the system,
as have multi- and bi-lateral development organizations. The
World Bank, the leading development agency in this respect,
has invested resources totalling about $2.4 billion in extension
activities.

  While T&V has been widely adopted, its progress has not
been without problems. Key aspects of the system as it was
initially described [1] and later updated [2] are frequently
misunderstood or ignored. Moreover, while the principles and
basic organization of the system are simple and appear
straightforward, they mask an inherently complex system. It
is this apparent simplicity that encourages adaptation and
change. This is welcome so long as the basic principles of the
system, and hence the intended functions of each of its
parts, are understood.

  This paper describes the fundamental principles of the
T&V system, discusses common misinterpretations in implementing the system, and concludes by describing an example
of the author's current work in extension in Burkina Faso.

FUNDAMENTALS OF THE T&V SYSTEM

The basic goal of the T&V system is to build a professional
extension service that is capable of assisting farmers in
raising agricultural production and/or income and of providing

---

*The views and opinions expressed in this chapter do not
necessarily reflect the position or policy of The World Bank
and no official endorsement should be inferred. This chapter
is prepared from notes by Michael Baxter, World Bank.

137

appropriate support to agricultural development. There can be no one system of extension suited to all farming communities. The variation in agro-ecological conditions, socio-economic environments and administrative structures is such that one system cannot be expected to suit all conditions. To be successful, the T&V system must be adapted to fit local conditions. However, the flexibility that enables successful adaptations to be made in the system does not allow for adaptation of its basic principles.

The fundamental principles of the T&V system include: (1) professionalism; (2) a single line of command; (3) concentration of effort; (4) time-bound work; (5) a field and farmer orientation; (6) regular and continuous training; and (7) two-way linkages between extension and research. Each of these is briefly discussed below.

By professionalism is meant an extension service that is professional in all senses. Extension must have close ties with scientific researchers in order to be able to formulate technical advice that will be useful to farmers. Extension workers must also be able to identify production constraints in the field and to either advise farmers how to overcome them or relay them accurately to research. To achieve this, extension workers at all levels need to be trained continuously to handle their particular responsibilities. Only with professionally trained staff will credibility for extension be built within the farming community. As well, extension workers must receive the basic physical and administrative resources and other support that are required to perform their professional functions.

Agricultural extension services to farmers must be unified under a single line of command within an appropriate ministry and department. Support is required from research and education facilities as well as from other agricultural services and government authorities, but all extension staff should be administratively and technically responsible to only one body. Moreover, the department within which the extension service is located should be responsible only for extension, notwithstanding the necessary linkages with other activities and organizations.

Another fundamental principle of T&V extension is that there should be a concentration of effort on individual component activities. Concentration of effort is a feature of all aspects of the system. Extension workers devote full effort to extension: they are not responsible for the supply of inputs, data collection, subsidy distribution and loan processing, or any other activity not directly related to extension. They should, of course, advise farmers on where and how to obtain inputs, subsidies, loans and markets, and so on, and keep extension management and others well informed of farmers' conditions. Involvement in non-extension activities dilutes the technical focus of extension and the direction and impact of extension operation. Just as extension is best done by a professional extension service, so are these functions best performed by specialized staff trained and able to work full-time in their specific fields.

Within an extension service, each staff member has a defined task: all efforts should go to performing that task. Similarly, attention in training sessions is concentrated on specific points, and extension-oriented research concentrates on key production constraints that are experienced by farmers. An agricultural extension service in any area initially focuses on a small number of important crops. As the service's expertise develops and appropriate technical and other support is available, it may gradually incorporate other important crops and other production activities of farmers (such as animal husbandry and farm forestry). In sum, concentration of effort means that the entire extension system is focused on bringing about the greatest and earliest possible increase in the production and income of the farmers it serves.

Time-bound work is another basic principle of T&V extension. Farmers must receive technical advice and assistance from extension agents in a regular and timely fashion so they can make the best use of resources at their command. The extension agent must visit his farmers on a fixed day that is known by all farmers and all other extension staff must make timely and regular field visits to fulfil their job functions. Technical Subject Matter Specialists (SMS) must discuss technical recommendations for a specific area and for particular farming conditions with research on a monthly basis, and subsequently teach extension agents on a frequent, regular (usually, fortnightly) basis. The regularity and frequency of field visits, research/extension workshops, and extension training sessions also ensure the opportunity of frequent feedback from the field to extension management, research and other agricultural services. Any break in this time-bound system of training, visits and feedback makes effective extension difficult.

Without a field and farmer orientation extension cannot be effective - to serve farmers, extension must be in contact with them. Moreover, extension's contact with farmers must be on a regularly scheduled basis, and directly with farmers representing all major farming conditions and socio-economic types of the broader farming community. In addition to frequent, regular contacts between extension agent and farmer, all other extension staff (from first-line supervisors to the service director), as well as researchers and others involved in agricultural services must have frequent exposure to farmers in their fields and villages. The ultimate test of extension's success is its impact upon farmers. Intelligent exposure to the field and farmers can quickly indicate the strength of all components of the extension system: whether

technical messages are appropriate (and hence whether training and feedback and indeed research are effective); whether
there is adequate supervision by each level of staff; whether
coordination with agricultural input organizations is satisfactory; and how extension field operations, administration and
training and feedback to agricultural services may be adjusted
for a better impact in the field.

To achieve the necessary field-and-farmer orientation,
one basic question should be kept in mind: how will any
proposed activity most readily benefit farmers? If the benefit
is unclear, or it can be achieved in a less complicated
fashion, there is need to rethink the approach. To ensure
that extension staff concentrate on effective farmer contact,
and in view of the generally unproductive outcome of reports,
the administrative and report-writing responsibilities of all
extension staff should be non-existent (or perhaps minimal).

Regular and continuous training of extension workers is
required to teach them the specific production recommendations to be discussed with farmers in the coming weeks,
and also to upgrade generally the professional skills of extension staff. The cumulative total of regular, frequent (and
practical) training can have a significant impact on the knowledge and ability of staff. Moreover, regular training sessions
are a scheduled venue for feedback between farmers and
agricultural services, and for extension staff to exchange
information and help them learn from each other's experience.
Without regular training, extension workers have very little
to say to farmers - and the so-called extension service has
little reason for functioning.

A final fundamental principle of the T&V system of
extension is that there must be close, two-way linkages
between agricultural extension and research. Problems faced
by farmers that cannot be resolved by extension field staff
and their SMSs must be quickly forwarded to research.
Improvements in technology developed by research must be
equally quickly fed into the extension system to be discussed
with farmers with the appropriate resources. Without the
technical content that comes from research, extension has
little to do in the long run. Similarly, without the orientation
to farmers' conditions and priorities and farmers' reactions to
recommended technologies that extension is able to collate on
a regular and representative basis, research cannot remain
effective for long. Research's awareness of a reaction to
actual farm conditions is increased through its responses to
problems that have been put forward by extension workers,
through the training of extension staff, and through field
visits: regular research-extension meetings (i.e. monthly
workshops) ensure this interaction takes place with sufficient
frequency to ensure impact.

Just as extension is unable to function in the long run
without close research support, so research depends on

extension for its ability to serve farmers effectively. This contributory role of extension to research explains why, even in the absence of an effective agricultural research system, it is important to develop a farmer-oriented extension system. Extension can provide the pressure to get an agricultural research system to focus on relevant farmer problems and to develop appropriate technology. Without such pressure, even a potentially responsive system as farming, systems research is likely to have a weak farmer orientation. In sum, in the absence of an effective research system, it is not that extension is not required - in such circumstances it is required even more.

These fundamental principles of T&V extension are straightforward. With them in mind, the operational components of the system are similarly easy to understand. Training and Visit activities of all staff adhere to fixed schedules to ensure the necessary field-and-farmer-orientation are adequate and appropriate during the staff training. Subject Matter Specialists contribute to research/extension linkages and the quality of extension's technical messages and to the service's overall professionalism. Monthly workshops are held to bring senior extension technical and management staff into direct contact with agricultural research, and to ensure research receives frequent direct feedback on farmers' reactions to recommended technology and to problems requiring further investigation. Field exposure by all staff helps ensure that the fundamental field and farmer orientation and the necessary quality and relevance of training and feedback to agricultural services (including research) are maintained.

While these principles are straightforward, experience in the field suggests they are not readily understood. Examples of such misunderstandings follow.

Fixed Work Schedules
Without work programs that are location- and time-specific, few people can work effectively. The physical circumstances under which extension operates - scattered staff who are only in infrequent direct contact with a supervisor and are responsible for large numbers of farmers over broad areas - makes a fixed, known program even more important. Without such a program, it is difficult to ensure that a worker fulfils his functions systematically (i.e. covers all required villages or, in the case of supervisors, workers in a systematic manner); and it is not possible to monitor and thus improve the quality of work. Equally important, farmers should know who is meant to work with them and when this work is meant to take place. Farmers' access to such knowledge facilitates the supervision of extension staff as farmers themselves take on a supervisory role. Farmers are eager to know who is meant to work for them and when - the main objection to

141

having fixed work programs usually comes from staff not willing to be tied down to monitorable responsibilities or from parties wanting to use staff for non-extension ends.

Contact Farmers

A second area where confusion is common is related to the concept of the contact farmer. Contact farmers are not another level of extension worker. They were devised for two main reasons. First, since an extension worker cannot (and does not need to) visit all farmers in his jurisdiction, contact farmers are a means of making his basic task of meeting farmers manageable. Secondly, through ensuring that the contact farmers of any particular extension group represent all major production and resource conditions of the farmers of that group, they are a means of seeing that extension confronts all such conditions. There are potential difficulties in the selection of contact farmers in that they may not be truly representative of the farming community from which they are selected and they may not be full-time or serious farmers. But in practice these problems can be overcome with careful selection of contact farmers and by monitoring their interest in extension activities.

The advocacy of contact farmers does not mean that extension cannot work with farmers' groups. Indeed, extension staff working with contact farmers should utilize group meetings to complement individual contacts, and other farmers should be encouraged to participate in the extension agent's discussions with contact farmers. In some situations, the farming community may be more inclined to work through groups rather than individual contact farmers. That being the case, care should be taken that the range of production and resource conditions of the community be adequately represented in the group and in extension activities. Moreover, extension field operations centered on groups do not preclude the need for work in individual farmers' fields and direct contact between extension staff and individual farmers. The basic point in the discussion of 'group' versus 'individual' approaches is that any method that enables effective farmer/ extension contact should be pursued - and that is unlikely to be either a pure 'individual' or 'group' approach.

Supervision and Leadership

The role of supervision and leadership in effective agricultural extension is another area of confusion. Some claim that 'The T&V system is well-conceived and organized - the only problem is that it requires leadership.' The only response to that assertion is 'name one administrative system that does not require leadership'.

Despite detailed descriptions of the T&V system's organ- ization and the supporting job descriptions, one of the weakest areas of any such system's implemention is likely to be supervision. Supervisors often do not supervise. They may be involved in a range of other functions that 'prevent' them from fulfilling their intended roles, they may lack the transport required for their supervisory function, or they may not do the work for other reasons. No matter the reason, unless staff are able and expected to supervise as their position requires, the T&V system (like any other extension system) will operate below its potential - indeed it will in essence not operate.

Subject Matter Specialists

A fourth area of confusion is the role of Subject Matter Specialists. One is tempted to say that SMSs are the most important component of the T&V system, though this is at odds with the fact that each component (extension agents, their supervisors, regular training and farm visits, research/ extension interaction, etc.) is equally important to the func- tioning of the system. It is enough to say that SMSs have a crucial role in the system, and that an extension service without them is unlikely to operate effectively.

Subject Matter Specialists are the prime trainers of extension agents and their immediate supervisors. They are responsible for checking the way in which recommended practices are being presented to farmer's in the field and monitoring farmers reactions to recommendations and the extent to which the recommendations cover the major crops and practices of farmers. SMSs are also the extension staff who are in most direct contact with research. They are responsible for ensuring that research is aware of farmers' conditions and technological needs and they work with research to identify production recommendations relevant to each set of important farming conditions. To fulfil these functions adequately, SMSs are expected to spend approxi- mately one-third of their time making field visits, one-third training other extension staff (primarily in fortnightly train- ing sessions) and one-third in contact with research (in extension/research workshops, contact with individual researchers in research libraries, and working on some experiments in conjunction with research workers). Clearly a 'T&V extension system' without SMSs is not 'T&V'.

BURKINA FASO: APPLICATION OF T&V PRINCIPLES

There are two basic forms of agricultural extension organ- izations in Burkina Faso. One is derived from the Training and Visit system and the other may be called (for Burkina

Faso) the 'traditional' extension system. The former is princi-
pally found in the Bank-assisted Brougouriba, Hauts Bassins
and Volta Noire ORDs [3] (districts).

In the T&V-derived system, as in T&V extension, staff
work largely on extension, are meant to visit their assigned
villages on a regular, rotational basis, and receive more
intensive training than under the traditional extension
system. There are, however, a number of key differences
between the T&V-derived system in operation in Burkina Faso
and 'T&V' extension elsewhere. In particular, the system in
Burkina Faso is without regular, two-way linkages with
research; there are no Subject Matter Specialists responsible
for the regular inservice training of extension staff or for
ensuring a regular, relevant and productive interaction with
research; extension field activities take place for only about
seven months of the year; visits to individual farmers are
made on up to six-week intervals; extension staff training
takes place only prior to the 'field season'; there is very
limited, if any, immediate feedback from extension workers to
agricultural research or other agricultural services; recommended technology is generally not adjusted for agro-
ecological, farmer resource or input availability conditions;
and supervisors are greatly constrained in the frequency and
quality of support they give to field staff. These conditions
were more or less common in Hauts Bassins, Brougouriba and
Volta Noire. In addition, in Volta Noire at least, perhaps only
one-third of field level extension staff time is devoted to
extension.

In the 'traditional' system in Burkina Faso, extension
staff may have a model timetable indicating their programmed
activity on particular week days. As these activities are not
linked to particular locations, however, the timetables cannot
be used as a basis of work planning or supervision. In
practice, the attention of extension staff is focused on a small
number of villages and activities, and a considerable amount
of time is spent on non-extension activities (such as acting as
storemen for input depots).

Undoubtedly some useful extension work takes place
under both T&V-derived and traditional systems. However,
even where adequate transport facilities and operating funds
are available (which is not common), the organization of
extension on the whole is such that its impact is not widespread and it is of limited replicability. Interestingly, extension field staff of both systems often indicate that the major
constraints to their work are (after limited transport)
inadequate training and insufficient guidance and involvement
in their work by senior staff.

A key ingredient in the analysis of the current extension
situation and the delineation of methodological requirements
has been active Burkinabe participation. The relevant central
government extension (DFOMR and SVAR) and research

(INERA) departments and the ORDs themselves have been actively involved in analyzing present extension systems, considering the linkages between extension operation, other rural services, and national priorities and objectives for rural development, and have been in developing extension systems adjusted to local resources, conditions and priorities. A pilot program to strengthen extension activities is being developed that takes account of the Government's desire to develop feasible Burkinabe initiatives for effective agricultural and rural development.

The extension methodology being experimented with under the pilot program is based on the following general principles:

- The structure of the extension service should be sufficiently flexible to respond to relevant local priorities, socio-economic conditions, available appropriate staff and other resources, and administrative structures.
- The extension methodology developed under the pilot program should be replicable, with local adjustment, on a national scale.
- Experience with the pilot program should indicate possible ways to strengthen other rural services to be more responsive to farmers' needs.
- Formal and informal village groups should have a central role in the organization of villagers to identify and prioritize their development objectives, and to undertake village-level development activities.

In addition to these general principles, the extension methodology is based on commonly accepted features of effective field extension services. These include, for example:

- The basic function of an extension service is to promote useful and remunerative technological change among farmers, and to keep agricultural research and other rural services well informed of farmers' conditions and needs.
- Extension staff should work full-time on extension activities which encompass all productive activities on a farmer's land (and which ultimately should be served by one extension service), though effort is initially concentrated on a small number of important activities.
- Extension staff should be expected and enabled to perform relevant, feasible and monitorable tasks.
- The basis of effective extension work is a well-trained staff who work closely with researchers, and who conduct their field and training activities in a regular, systematic function.
- Farmers should be closely involved in identifying their

technological needs and monitoring the work of extension agents.
- Recommended technology should be adjusted to local agro-ecological and cultural conditions, and enable farmers to make the best use of their available resources.
- As an extension worker cannot deal effectively with all farmers in his jurisdiction, complementary methods of mass (group meetings, collective fields, coordinated radio programs) and individual contact should be utilized.

The extension methodology that is being implemented under the pilot program takes account of both the general principles of extension reorganization and the commonly accepted features of effective field extension services as noted above. Some main features of this methodology are:

- A close, two-way linkage with research.
- Establishment of Subject Matter Specialist teams within the extension service to liaise with and be trained by research, and to train extension field staff.
- Regular training of SMSs (by research) and field staff (by SMSs).
- Extension field activities take place all year on a two-week cycle, with both group and individual contacts being handled on the same day.
- Village groups and their collective fields are the initial point of contact between extension field staff and farmers.
- Supervisors make frequent systematically scheduled field visits.
- Locally-relevant key technical messages for each area are promoted both on collective fields and in small areas on individual farmers' fields.
- Animal husbandry (silage, fodder crops, organic manure, draft animals, etc.), soil and water management and reforestation will be part of the responsibility of extension field staff, and are integrated into their training and research support.
- Farm trials (now being conducted by semi-autonomous adaptive research centers) are integrated into extension and research work.

While these basic characteristics are common to extension organization in all ORDs participating in the pilot project, there are some differences in practice among them. For example, there are slight differences in supervisory structures and responsibilities, work in some areas is based more on villages than on village groups, and the importance given to livestock, soil and water management, and reforestation varies between ORDs. The role of non-governmental organizations' activities and training also differs between ORDs.

By its very nature, the pilot program to strengthen agricultural extension is experimental. The prime concern in 1986 is to determine the broad outlines of an effective agricultural extension service. Evaluation of 1986 experiences will be a prerequisite to determining the form of extension organization in the future. Simultaneously, attention has gone to how rural services in general may be organized more effectively. Just as initial focus is on getting one service - extension - operating effectively, so should subsequent involvement in other services be on a gradual prioritized basis.

Village groups have an important role in agricultural and rural development in Burkina Faso. Experience with formal and informal village groups is an important consideration in the handling of other rural services. Village groups are the initial and central point of contact for extension workers, and provide (through their collective fields) an important venue for the introduction and demonstration of recommended technology. Strong village groups will greatly enhance the relevance and impact of extension work, as much as of all other rural services. Since some groups are exclusively composed of women, they also offer an opportunity for extension (and research) to confront systematically the requirements of women as farmers. To have a significant impact on development, however, groups usually require encouragement and, initially, external guidance [4]. An important issue to be resolved, if possible in conjunction with the pilot program, is how this should be done.

Closely related to the operation of both the extension service and village groups are the Centres de Formation des Jeunes Agriculteurs (CFJAs). There are almost as many CFJAs as there are extension agents in Burkina. Often CFJAs function as substitute primary schools rather than as centers of training for older teenagers in agricultural skills as intended. Given the evident need for 'animateurs' (to work with village groups to improve the organizational and analytical skills of village groups, and help ensure that they productively utilize governmental services) and for an accelerated program to enhance the skills of village group leaders in the context of the adult literacy program, it would perhaps be more productive for CFJA staff to concentrate on these areas. This is an example of the type of issue that needs resolution once an effective extension service is established.

Aside from establishing an effective agricultural extension service, Government has indicated that the strengthening of the agricultural credit, input delivery and marketing systems, as well as of the adult literacy program, are priorities. Since village groups already have an important role in credit, input delivery and marketing, the development of village groups parallel to the establishment of an effective extension service takes on greater importance. Greater emphasis by Government on natural resource development, which involves village level problem identification and an agricultural extension service active in soil and water management, reafforestation and livestock development, adds even more to the need for strong village group organization.

A primary constraint to effective extension work under the Burkina Faso pilot program may be inadequate operational support available for extension staff at each level of supervisors and that supervisors and technical specialists therefore do not perform their required functions. Other prerequisites for effective extension work will be the gradual upgrading of staff through intensive in-service training, the continuing development of a farmer-oriented research system, and the creation of effective village groups. While attention will necessarily go in the first instance to establishing an effective field extension service, the resource requirements of all rural services should be seen in totality. When determining how to adopt and improve extension methodology more widely in Burkina, care must be given to ensure that there is minimal duplication between rural services and that the system adopted (which may vary regionally) is the most simple and efficient way to meet national development objectives.

Our work in Burkina Faso shows how easy it is for a 'T&V' extension system to lose sight of the fundamental principles of the system, and for a far-ranging restructuring of the service to be necessary. This experience is not uncommon. It is for this reason, that this paper asserts that work in the second decade of T&V should be with the motto 'Back to Basics'.

NOTES

1. World Bank, (1977) Agricultural Extension: The Training and Visit System.
2. Benor, D., Harrison, J.Q., and Baxter, M. (1984) Agricultural Extension: The Training and Visit System, Washington, DC: World Bank.
3. An ORD (Organisme Rural de Developpement) is essentially a district.
4. Jean Morize presents in L'Animation des Groupements Villageois (Editions Forhom, Paris, 1985) a guide on how village groups might be organized so that their developmental needs are identified and prioritized, and they are able to deal effectively with governmental agents (such as extension workers).

Chapter Ten

MAKING EXTENSION EFFECTIVE IN KENYA:
THE DISTRICT FOCUS FOR RURAL DEVELOPMENT

Christopher A. Onyango
Egerton College

BACKGROUND

Kenya, a former British colony located in East Africa, covers
an area of 225,000 sq.km. [1] with a population of 20 million
people [2]. Eighty per cent of the population of Kenya is
engaged in some form of agriculture and/or related industry,
freeing 15-20 per cent for manufacturing. Although a tropical
country, the prevailing climate ranges from warm tropical
along the Indian Ocean coastline and its hinterland, to cool
temperate Afroalpine-like in the central highlands (snow
covered Mt. Kenya region) and western regions (Mt. Elgon
region). The northern and north-eastern region occupies
two-thirds of the country and is a barren semi-desert land.
Kenya's agricultural development is concentrated in only
one-third of the country.

Like most other developing countries in the tropics,
Kenya's agriculture can be divided into: (1) small subsistence
mixed farming (6-50 ha) and (2) large scale farming, including plantation agriculture (50-2,500 ha. or more) [3]. Farm
land is not easily available due to population pressure and
urbanization on arable land. Existing large scale farms are
subdivided into smaller units when necessary. This system
generally reduces land productivity, but ensures that small
farmers have an area for subsistence through careful planning
of production systems. The Government of Kenya creates an
awareness of the need for maximization of the available agricultural resources to the population through agricultural
education. Agriculture is a compulsory course in primary,
secondary schools as well as Teacher Training Colleges. The
adult population is reached through radio, television, films,
farmers' journals, annual agricultural shows, field days for
research stations and agro-chemical companies, and farmer
training centers.

Manpower training in agriculture has been achieved
locally through several universities: The University of
Nairobi, Moi University-Eldoret, University of Eastern Africa

149

MAKING EXTENSION EFFECTIVE IN KENYA

at Baraton, Kenyatta University and Egerton College. Individuals have also been trained in the universities of the neighboring nations, Europe, Asia and USA. Most of the graduates find employment opportunities in extension, research and private agro-based companies.

Government research stations, as well as those of private companies which are distributed throughout all ecological zones, play a vital role in the advancement of the agricultural sciences. Establishment of national as well as international agricultural research institutions has further strengthened Kenya's agricultural capabilities, making her a country which offers extended manpower training to about 25 additional nations of Africa. FAO reports on extension systems [4] and manpower assessment for trained agricultural workers in Africa [5] show similarities between Kenya and other situations in Africa in these areas. However, significant changes have been made recently in the policy, structure and management of extension to increase its effectiveness in Kenya.

POLICY FOR DEVELOPMENT

Development strategy in Kenya has shifted markedly since 1984 from a centralized focus, with planning and implementation of development activities directed from national offices, to a more decentralized system where most of the work is done at the district level. This strategy, known as the 'District Focus for Rural Development' is based on a complementary relationship between ministries and districts. According to the official government policy paper [6], the objective of this approach is 'to broaden the base of rural development and encourage local initiative in order to improve problem identification, resource mobilization and project implementation'.

There are clear lines of operation drawn between the ministries and the districts. The ministries are responsible for coordinating multi-district and national programs such as research stations, provincial hospitals and major roads. The districts, on the other hand, deal with specific projects of relevance to the particular areas such as rural health centers, cattle dips and village water supplies. Thus, where the activities coordinated by the ministries serve a broader population, those managed by the district serve the local target population.

ORGANIZATION OF THE DISTRICT AND ITS
RESPONSIBILITIES

The District Commissioner is the Chief Executive Officer for district rural development activities. He is the chairman of

the District Development Committee (DDC) which is responsible for rural development planning, coordination, project
implementation, management and development of resources,
overseeing local procurement of goods and services, management of personnel and provision of public information.

The District Development Committee is composed of civil
servants, elected leaders at the national and local levels, and
representatives of non-governmental organizations (NGOs)
including representatives of self-help groups and donor
organizations.

The main responsibility of the DDC is to develop District
Development Plans (DDP) that are well documented, prioritized and funded with available resources. The plans are
also expected to be cost effective and consistent with broad
national policies. The DDC also prepares annual work plans to
coordinate the implementation of projects, especially those
which require interdepartmental cooperation.

Funds for projects developed and implemented by the
DDCs are controlled at the District level through the District
Treasury. This ensures that project implementing officers who
receive Authority to Incur Expenditure (AIE) can release
funds without delay.

The District Tender Board (DTB) is authorized to spend
up to Kshs. 60,000 for purchases of equipment and supplies
for local projects and to ensure the monies are spent according to specific projects. The budget cycle of the DDC is
illustrated in Figure 10.1.

RESOURCES FOR RURAL DEVELOPMENT

There are several sources of funding for DDC-approved rural
development projects. These include (1) individual ministries,
which provide funds for specific high priority projects (2)
combinations of ministries, which provide funding for multi-
district projects such as provincial hospitals, provincial roads
etc. (3) local authorities (e.g. County Councils and Municipal
Councils) which provide funding for projects within the
authority but with the approval of the DDCs (4) self-help
groups which provide resources in the form of money,
materials and labor for locally desirable projects (5) non-
governmental organizations which support various special
programs of significance in rural development.

In order to ensure that the District Focus strategy
works successfully, an Inter-Ministry Coordinating Committee
has been set up. This Committee consists of permanent secretaries for Development Coordination, Finance, and Planning,
and a Directorate of Personnel Management and Provincial
Administration [8]. The Committee operationalizes the strategy
throughout the country.

151

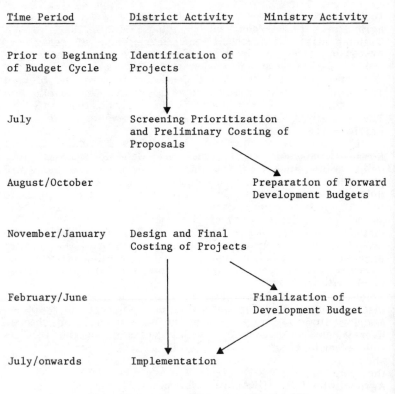

Time Period	District Activity	Ministry Activity
Prior to Beginning of Budget Cycle	Identification of Projects	
July	Screening Prioritization and Preliminary Costing of Proposals	
August/October		Preparation of Forward Development Budgets
November/January	Design and Final Costing of Projects	
February/June		Finalization of Development Budget
July/onwards	Implementation	

Fig. 10.1:   Annual budget cycle [7]

At the lower levels, work is supervised by a Provincial Monitoring and Evaluation Committee. This Committee is composed of the Provincial Commissioner, Provincial Heads of Government Departments, members of the Cabinet resident in the Province and individuals providing special expertise. The group tours all the Districts in the Province, visits projects and holds discussions with implementing groups including the DCs and the DDCs. It is their role to advise, critique, and assist in the overall implementation and performance of both District-specific and multi-District projects.

The strategy of using Districts as the focal point for development planning and implementation is a challenge to development workers in Kenya at all levels. A government circular (1985) issued by the Chief Secretary spells out procedures and interrelationships which must be carried out to ensure that the system works [9].

Agricultural activities, especially research and extension, are expected to function within the framework of the District

Focus structure. The Kenyan economy is based on primary agricultural production and the country has adopted the Training and Visit (T&V) system model as its basic framework for disseminating agricultural knowledge to farmers.

T&V AND THE DISTRICT FOCUS STRATEGY

The basic T&V structure, as articulated by Benor and Harrison [10], emphasizes continual contact between subject matter specialists (SMS), the extension agent (VEW) and the farmers' group (either through a contact farmer or directly) on a fortnightly sequence. It requires the existence of planned information packages to be delivered to the farmers during regular meetings.

The T&V approach was first introduced in Kenya in 1982 as a pilot project in two districts. Maize yields in these districts were subsequently 42 per cent above normal, which proved to be the main catalyst for expanding the project. Phase II added eight districts in 1983. Phase III added ten districts in 1984, and Phase IV added ten districts in 1985. Kenya now has 30 districts using the T&V approach, all of which are arable mixed farming areas. The remaining eleven districts are predominantly pastoral. They follow a different model in their agricultural strategy, with emphasis placed on livestock population development, control and marketing.

According to the T&V model each District in Kenya should have its own extension program, planned according to the activities of the farmers in the area. The District Agricultural Committee (DAC) is normally expected to determine agricultural development priorities in a district. The DAC is composed of selected local farmers, agricultural extension workers at the district level, a district commissioner, representatives of the planning section, as well as researchers. The DAC, therefore, functions as a technical sub-committee of the DDC. Its decisions must be ratified by the DDC, especially where the development of projects is concerned.

The implementation of an extension program in a district begins with the development of a comprehensive cropping calendar by the district agricultural staff. There may be differences among the various divisions in a district or within various ecological zones. This variation has to be accounted for during the planning stage. In the planning of the cropping calendar, all activities must be clearly specified.

The planning stage is followed by the creation of a fortnightly training schedule and monthly workshops. During planning sessions, the frontline workers, the SMS and researchers review program activities, progress and constraints. The incorporation of livestock activities has been a major difficulty under this approach since livestock farming does not fall in any well defined pattern like crop farming. However, within this framework, efforts to plan livestock activities have been successful to a certain extent. Activities such as grazing control, dipping programs, tick control, disease control, fodder development, calf rearing and clean milk production and handling have been successfully incorporated into the T&V system.

The National Extension Program (NEP), as the T&V approach is generally referred to in Kenya, emphasizes extension activities at the local levels, particularly the district level. In this case extension activities are coordinated with the District Focus Plans. In addition, the Ministry of Agriculture and Livestock Development (MOALD) has set up a program of monitoring and evaluating extension activities in the field. The unit of the ministry which does this is expected to gather data and provide monthly feedback to the District Agricultural Officer. In turn, the District Agricultural Officer identifies weak spots in the district program and makes changes as necessary.

Since its inception the T&V approach in Kenya has resulted in the following achievements [11]:

1. A uniform management system has been established from Headquarters down to the farmer level.
2. Staff members perform their duties according to clearly defined job responsibilities.
3. Improved extension research linkages have been established enabling the smooth transfer of available technology, as well as joint planning, implementation and evaluation of on-farm trials.
4. Regular monthly and fortnightly staff training programs have been implemented, focusing on practical skill training.
5. Farmers are willingly cooperating with extension staff without receiving free inputs.
6. Existing women's group are participating in the program and are playing an important role in the dissemination of newly recommended practices.
7. Chiefs and other local administrative staff participate actively in the program and are helping to achieve its objectives.
8. Constructive pressure from farmers has been brought to bear on research for new relevant technology, as well as on supporting agencies dealing with credit, inputs, and marketing for better service.
9. The actual visual impact of the program can be seen on farmers fields showing adoption of the various newly recommended practices.
10. Field staff have seen that their activities have brought substantial improvements and they have become highly motivated.

These achievements indicate that a concentrated and successful effort is being placed on extension work at the local level in Kenya. The combined strategies of District Focus for Rural Development and the T&V system have created an atmosphere where farmers, directly or through their representatives, can contribute to the planning and execution of agricultural activities. It has also contributed to a situation where scientists, administrators and other agents of change have access to and continued contact with farmers and their problems giving the farmer an opportunity to provide remedies.

INCENTIVE SYSTEMS IN EXTENSION

The effectiveness of agricultural extension and indeed of agricultural production is highly dependent on the associated incentives. These incentives relate both to extension workers and to farmers. Sources of incentives vary depending on the resources within a country, the projects being undertaken and their utility to those who are likely to provide the particular incentives. These sources are likely to be government or the private sector.

Government Incentives [12]
The government provides incentives to extension workers in many respects. These include:

1. Salary Structure
The salary structure for extension workers is comparable to that of other government employees. Extension workers are placed in relevant job groups according to their training and qualifications, as are all other civil servants without any discrimination. These positions are from the lowest hands-on-practical level to the highest managerial and administrative levels.

2. Professional Career Structure
Professional career structure is closely connected with salary structure. Opportunities for further training exist for extension workers who desire and are capable of improving their knowledge. Such individuals are posted to positions of greater responsibility upon completion of their training. Professional staff are thus able to move vertically or horizontally to more challenging jobs in other sections of the organization.

3. Allowances

Other incentives given to extension workers by the government include free housing or house allowance; leave or leave allowance; maternity leave of up to two months for married women; and loans and advances for purchase of such items as land, house, motor vehicles, furniture and other household items. In addition they are given special allowances for various special duties and responsibilities. Upon retirement from the service, they are entitled to a pension for which they are not required to make a contribution.

4. Other Government Incentives

a) <u>Inputs and Guarantees</u>: The incentives given to the farmers by the government are dependent on government needs. Apart from providing a well organized and adequately staffed extension service, the government gives farmers loans for investment in various production activities. These loans are usually provided in the form of services such as land preparation (ploughing and harrowing), provision of seed and fertilizer, and provision of water in case of irrigated crops. In special projects, the government provides transport for produce as well as marketing channels. The farmers are also guaranteed a return on their investment through the Guaranteed Minimum Return (GMR) schedule on short term crops such as maize and wheat. This ensures that even if there is crop failure due to drought or wildlife damage, the individual farmers can claim a minimum refund on their investment.

b) <u>Price Reviews</u>: Continual review of agricultural commodity prices is a way of stabilizing the consumer market in the country, and also a method by which farmers are compensated adequately for their efforts. Crops which are reviewed regularly include maize, wheat, sugar cane, beans and cotton. In addition raw milk prices are adjusted frequently in order to adjust for drought and/or rainy weather conditions. These incentives motivate farmers to produce at optimum levels.

Incentives by Parastatals

Parastatal organizations, which are semi-autonomous government institutions, also provide incentives to their extension workers. Kenya has several parastatal organizations which are responsible for agricultural production and employ their own extension workers. They include organizations such as the Kenya Tea Development Authority (KTDA), the Pyrethrum Board of Kenya, the Coffee Board of Kenya, Kenya Seed Company, and various commodity organizations.

The Kenya Tea Development Authority provides planting material, fertilizers and an efficient extension service. It has a marketing structure which recognizes the quality and quantity of the farmers product and compensates them accordingly.

The Coffee Board of Kenya provides fertilizers, coffee berry disease spray and extension advisors. It has a marketing structure which responds to the international coffee market situation and compensates farmers well.

The Cotton Board provides seed, fertilizer and sprays to farmers. Farmers are paid upon delivery of their cotton lint and prices are reviewed regularly by the government and adjusted so that they are beneficial to the farmers.

The Pyrethrum Board acts similarly and provides planting material, fertilizers, extension advice and a stable market. It is also sensitive to the international demand for natural pyrethrum and compensates farmers according to the market situation.

The various sugar companies in the country are encouraged to produce enough sugar to meet national needs. In turn, they provide their outgrower farmers with various incentives. These include land preparation, planting material (seed cane), weed and pest control, harvesting and transport. Arrangements are made to pay the farmers when their sugar can is delivered to the factory.

The above mentioned organizations and other public sector institutions ensure that production levels in the major agricultural commodities are maintained. They seek to satisfy local demands for food and other raw materials, as well as the demand for export. Some of the organizations may not have an active extension section and instead depend on government workers. Those with an active extension branch, however, ensure that they have regular contact with farmers. It is through this regular contact that they improve the effectiveness of extension.

Private Sector Incentives
The private sector plays an important role in the effectiveness of extension in Kenya through direct involvement in farming activities. Relevant organizations in the private sector include: farmer cooperative societies, banks, private agribusinesses and industry.

1. Cooperative Societies
Cooperative societies are formed and operated by farmers themselves. They provide a medium through which farmers can obtain credit, credit-guarantees, reduce input costs such as fertilizers, seeds, herbicides, insecticides, farm machinery and implements. Cooperatives which deal in marketing also

guarantee stable and competitive prices. Among the most prominent of such organizations are the Kenya Grain Growers Cooperative Union (KGGCU), the Kenya Cooperative Creameries (KCC), and the Kenya Planters Cooperative Union (KPCU). All three organizations are highly production and market oriented and provide the central lobby to the government on behalf of the farmers. The cooperative societies do not normally engage in direct extension contact with farmers. However, during the agricultural shows at the local, district, provincial and national levels they demonstrate most of their activities for the benefit of the farming public.

2. Banks

The role of banks in making extension effective is determined primarily by their own profit requirements and lending policies. International funding for major agricultural projects such as cotton production, coffee improvement and crop diversification is channelled through commercial banks. The banks also provide limited loans for farmers who can provide some type of security (usually land title deeds). Such loans are used to purchase land or equipment and are normally short term. Long term loans can be obtained mainly from the Agricultural Finance Cooperative (AFC) or the Cooperative Bank.

Although it is traditional for the banks in the country not to engage directly in farming activities, the emphasis on rural development through the district focus is making it necessary for banks to develop a rural outlook. As a result, several financial institutions have sprung up and most of them are showing a greater willingness to give credit for rural development activities, including farming.

3. Private Business

Private business is another area which supports the effectiveness of extension in Kenya. There are many agri-businesses which sell seed, fertilizers, pesticides, herbicides, acaricides, etc. to the farming community. Agri-businesses also buy raw materials from the farmers either for direct sale elsewhere or for processing. Most of these businesses support their activities with extension advice. The most prominent among these businesses are Welcome Kenya Limited, Twiga Chemicals Ltd., Hoechst E.A. Ltd., Pfizer Ltd., Ciba Geigy and Shell Ltd. In addition there are various smaller organizations which deal with specific crops or products.

Representatives of these organizations provide extension advice to their clients, and engage extension staff in seminars and workshops to demonstrate products and their uses. From such seminars specific recommendations are formulated and standardized for general application in extension services.

Private businesses also engage in promotional activity through radio, television, newspapers and magazines.

4. Industry
Industry plays a key role in improving the quality of extension work in Kenya. The Kenya Breweries, for example, employ a large team of qualified agriculturalists and deploy them to provide technical advice to farmers with whom the organization has barley growing contracts. Similarly the British American Tobacco (BAT) Kenya Limited deploys staff to work with farmers in tobacco growing areas. Others include: East African Industries and Oil Crop Development Limited who together promote oil crop growing (especially sunflower) as raw material for oil based products such as soap and cooking oil; Kenya Canners who promote pineapple growing; various sugar factories which deploy their staff to assist sugar farmers; and Kenya Orchards Limited (KOL) who promote the production of vegetables, fruits and pulses for canning.

As the emphasis for rural development continues, financiers and entrepreneurs are being encouraged to locate small scale industries in the rural areas. Consequently small scale private agri-based industries are beginning to emerge in the rural areas. These industries now tend to contract farmers for raw materials and in the process provide basic extension advice. A typical example is the Njoro Canning Factory which contracts with farmers to grow French beans for canning and export. The factory provides seed, fertilizers and pesticides. When the crop is mature and delivered, the farmer is paid promptly. This incentive system has helped to sustain a steady supply of the beans to the factory.

5. Voluntary Organizations
For many years voluntary organizations have provided complementary support to government effort in various aspects of development - particularly in education, health and social services. This support has been gradually extended to other areas of activity, including agriculture.

Church groups (i.e. National Christian Council of Kenya, the Salvation Army, the Catholic Relief Fund) have actively supported farmers' training and some have established Farmers' Training Centers (FTCs). They also have supported youth training by initiating the development of Village Polytechnics.

Other voluntary bodies such as Action Aid, World Vision and the National Greenbelt Movement also support extension efforts. Action Aid gives material support to youth in schools located in less developed areas who are encouraged to engage in simple agricultural practices such as vegetable growing, poultry and rabbit keeping through the guidance of school teachers and local agricultural staff. World Vision supports local voluntary groups to engage in agricultural activities that help to improve nutrition and raise general living standards.

Many voluntary organizations engage specialist extension workers to supervise their efforts and financial contributions. These workers collaborate with government extension workers at the operational level.

FARMER PARTICIPATION IN EXTENSION

The participation of farmers in extension activities has been a hallmark of agricultural development in Kenya. Participation is through activities organized either by the extension services or the farmers themselves.

1. Extension-organized Participation
Extension service workers organize participation by farmers in various activities. First, they organize training programs for the farmers in Farmers' Training Centers (FTC). There are FTCs or Rural Training Centers (RTCs) located in most Districts. These institutions are for training adults in various farming activities. The training sessions last one or two weeks and are heavily subsidized through government funding. During the training courses, farmers learn about specific topics in agriculture. The topics are selected by the FTC staff in consultation with the District Agricultural field services staff who in turn recruit the farmers for the courses.

Other extension-organized participation includes farm demonstrations (either in individual farmers' fields and homes or in research stations); farm tours to selected areas and/or other districts; and meetings organized by local extension staff or chiefs.

2. Farmer-organized Participation
For many years, farmers in Kenya have participated in the Agricultural Society of Kenya (ASK) organized annual shows. At these shows farmers display their farm and dairy produce, and show their livestock including cattle, sheep, goats, poultry and rabbits. In addition, they learn new practices, new techniques and new farming ideas from public sector organizations and other farmers.

Participation by farmers is generally intended to improve their knowledge about the latest agricultural practices. It is also intended to enable the extension workers to get feedback regarding the problems which farmers face. This interactional process improves the quality of extension plans, programs and activities.

SUMMARY AND CONCLUSIONS

It is not possible to have an efficient extension system unless several factors operate correctly. Such factors include the structure of extension for providing administrative and technical support, the organizational structure of public services which control the development strategy, the incentive systems for both extension workers and the clients of extension and finally the whole philosophy of extension and how it is expected to contribute to the national development effort.

In Kenya, the system of extension followed for many years has been what Rivera (1985) [13] refers to as a Typical Developing Country System, characterized by Ministry of Agriculture control, separate from teaching and research, highly centralized, and dealing with pre-determined programs. This approach has been found to be constrained in many respects and did not keep pace with development trends in the country.

The change from the 'typical' system to the Training and Visit (T&V) model was begun in 1981 and completed in 1985. Results of this new approach, as reported earlier, were encouraging enough to allow for its total implementation throughout the country. At the same time the government development strategy changed significantly to allow for development decision making at the grassroots level (in this case the Division and District level) and to encourage greater participation by local people. The mix of these two approaches is expected to yield improvements in development.

The incentive system may include personal gain to those concerned with extension or other forms of support (logistical, financial, administrative, etc.) Incentives have been provided by varied sources and to diverse recipients over the years. As the need to intensify agricultural production becomes paramount one expects that this aspect will continue to be stressed.

Extension is a change process. When it undergoes strategic, philosophical and management changes such as through the implementation of both the 'District Focus for Rural Development' and the T&V approach, caution must be taken not to expect spectacular changes too soon. Kenyans and others who promote extension and who wish it to be an effective medium for development must be prepared to wait patiently as the new structure evolves and stabilizes.

NOTES

1. Republic of Kenya (1972) National Geographical Map. Nairobi: Government Printer.
2. Republic of Kenya (1980) National Census Report. Nairobi: Government Printer.
3. Omamo, W.O. (1984) The Role of Training Institutions in Agricultural Management (unpublished), (speech read at Conference on Management and Training) Mombasa.
4. FAO (1984) Agricultural Extension Systems in Some African and Asian Countries. Rome.
5. FAO (1984) Trained Agricultural Manpower Assessment in Africa. Rome.
6. Republic of Kenya (1984) District Focus for Rural Development. Nairobi: Government Printer.
7. Ibid, p. 10.
8. Ibid, p. 15.
9. Republic of Kenya (1985) District Focus Circular No. 1/85, August. Nairobi: Government Printer.
10. Benor, D. and Harrison, J.Q. (1977) Agricultural Extension; The Training and Visit System. Washington, DC: The World Bank.
11. Republic of Kenya, Department of Agriculture (MOALD) (1984) African Workshops on Agricultural Extension and Research Linkages.
12. Republic of Kenya (1985) Review of Terms of Service for Civil Servants. Nairobi: Government Printer.
13. Rivera, W.M. (1986) Comparative Extension: The CES, TES, T&V and FSR/D. College Park, MD: The University of Maryland, Center for International Extension Development.

Chapter Eleven

THE ISRAELI EXPERIENCE IN AGRICULTURAL EXTENSION
AND ITS APPLICATION TO DEVELOPING COUNTRIES

Abraham Blum
The Hebrew University of Jerusalem
Faculty of Agriculture, Rehovot

THE PARADOX

Israeli agricultural development has been quite unique: it has
been rapid, has changed direction several times within one
generation, and has been highly science-based and capital
intensive. These phenomena are atypical of, or even contrary
to agricultural development in the Third World. How, then,
can Israel apply its experience to agricultural development in
countries with a long tradition of subsistence agriculture,
high rates of illiteracy and lack of capital?

Obviously, there is no possibility to emulate somewhere
else the solutions which were found to suit Israeli conditions.
This is already an advantage. Where developers have tried to
transfer the development blueprint from one country to
another, they have failed. Therefore, we should not be
content with a study of what changed, but should analyze the
role of extension in this process and extract generalizations
which can be transferred to other cases.

Accordingly, this discussion first outlines the main
phases of agricultural development in Israel. It then explores
the role of extension in this process and points out those
insights of value for extension work in other developing
countries.

AGRICULTURAL DEVELOPMENT IN ISRAEL

The intensive growth of Israeli agriculture can be roughly
divided into three decades, starting with 1950 following the
war of liberation. The first decade was marked by two major
trends:

a) A transition from a lack of fresh food to food
    surpluses;
b) Increased agricultural production by newcomers to
    Israel and to agriculture.

163

Until the establishment of the State of Israel (1948), agriculture in the area was quite extensive but not intensive enough for export purposes with the exception of citrus crops. The irrigated area was small. Much fresh food was imported from neighboring countries. This importing stopped after Israel became independent.

At that time, a flood of new immigrants reached the shores, mainly remnants of the holocaust and refugees. Within a few years, the population of Israel doubled - 400 new settlements were established within a period of three years. In contrast to the past, most of these settlements were Moshavim and not Kibbutzim.* Most new immigrants did not like the idealistic Kibbutz, in which not only production, but even consumption is arranged in a collective manner. They preferred the lesser degree of agricultural cooperation in the form of the Moshav. A 'compromise' form, the Mesek Shitufi, in which production is organized as in the Kibbutz, but each family receives a budget to cover consumption expenditures, did not become popular. Actually, over time both Kibbutzim and Moshavim integrated some ideas of the Meshek shitufi into their thinking as they adapted to changing situations.

Few of these new settlers had done any agricultural work. To introduce these refugees to farming demanded an extraordinary and new approach to extension. The results were not less extraordinary. Within six to seven years food production, which had started at a threshold, became so bountiful with fresh food surpluses that production quotas had to be imposed on all fresh products. By that time most available land was cultivated and all possible water resources were used for irrigation (See Figure 11.1). Water quotas also had to be imposed. Thus, for the 1960s new ways had to be

Table 11.1: Degrees of cooperation in various village types

| Village Type | Common (Cooperative) | | |
	Purchase of Inputs & Marketing	Production	Consumption
Kibbutz	+	+	+
Meshek (Moshav) Shitufi	+	+	-
Moshav Ovdim	+	-	-

*Moshavim and Kibbutzim are the plural form of Moshav and Kibbutz. Moshav is a short form for Moshav Ovdim (Workers' Settlement).

found to enable further agricultural development. Again two interconnected major trends marked the decade:

a) From mixed farming to specialization;
b) From local production only to an emphasis on export.

These trends again demanded changes in planning and in farming methods. New crops suitable for export (e.g. cotton) were grown, acreage was enlarged, and yields reached world records. The first 280 hectares of commercial cotton fields were planted in 1954. The average yield of a ton of cotton lints per hectare was promising, and within a decade grew by 1500 per cent. Today's average is 1.5 ton fibres and success-ful farms reach yields of two tons of high quality fibre per hectare. Some attempts, e.g. growing and exporting money-maker tomatoes, failed. Other crops were tried (e.g. sugar-beets and groundnuts) but did not compete with fresh produce on the market. They were less successful than cotton.

Quotas were fixed for fresh produce according to pre-ferred geographical areas and farmers became more special-ized. Specialization was particularly high in seasonal export crops which could reach the market at an optimal timing (after new techniques were developed and accepted by farmers). The major technique was growing strawberries, winter vegetables and flowers under plastic cover for export to Europe.

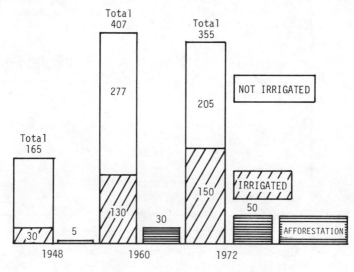

Figure 11.1: Land use (1000 ha)

The third decade, the 1970s, was marked by further changes which continue to date:

a) New techniques, especially to save water and man-power were developed;
b) New intensive crops, especially for export, were introduced;
c) The rural sector became increasingly industrialized.

By 1964 the National Water Carrier, which connected the already existing regional water scheme was operating. It enabled transport of Jordan water from the North of the country to the semi-arid Negev. But soon, the additional water resources were used up by further intensification of agriculture and by growing industrial and household needs. Thus the focus turned to saving water. The challenge was how to achieve higher yields with less water. The solution (or at least a breakthrough) is well known: drip or trickle irrigation. Again, a concentrated effort of research, extension and administration was needed.

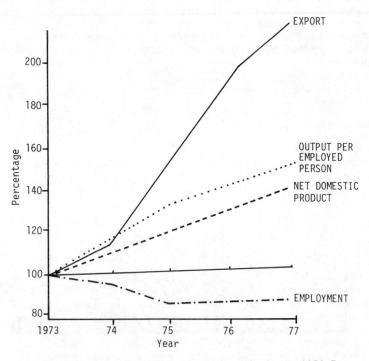

Figure 11.2: Trends in Israeli agriculture 1973-7

Further developments in greenhouse techniques and in controlled environments enabled reduction of water consumption even further. As well, the oil crisis of the 1970s hit energy intensive crops hard. For instance, rose growers which just had overcome many obstacles by using new technologies, such as plant hormones, and heating, now found themselves in an economic crisis. Prices in the export market no longer covered the investment in labor and capital.

Because agriculture could no longer be expanded, the rural sector turned to industrialization. The first to venture into this new realm were the Kibbutzim. The less close knit Moshav had and still has more difficulties in the industrialization process. Again, new solutions have to be found.

Before going into the development of agricultural extension, I shall look to the future of Israeli agriculture and to the challenges which lie ahead. No doubt some of the agrotechnical problems will remain with us. Some export crops like avocado and certain flowers have already lost their monopoly in the market. New crops and better adapted varieties have to be found, tried and grown commercially.

Techniques which are now being developed will become more important, e.g. better-controlled environments, tissue-cultures and products of bio-engineering. The export of seed materials and of knowledge will grow. Solar energy and the use of brackish water in growing halophytic crops will come into the foreground. But perhaps the most important changes will be of an economic nature. The Moshav structure is already changing. We shall look more at the family farm as a whole and at rural industrial enterprises and their place in an integrated village economy. All these changes will demand new approaches to extension work, which has already undergone important changes, adapting each time to new challenges.

THE DEVELOPMENT OF AGRICULTURAL EXTENSION

Until the establishment of the State of Israel, agricultural extension was in the hands of the British, who had established some experimental and demonstration farms, and in the hands of Jewish cooperative organizations (mainly Kibbutzim) who organized courses and evening lectures and published periodicals for their members.

This approach could no longer work with the new wave of immigration. Out of 400 new settlements established in the beginning of the 1950s, 250 were settled by new immigrants who had had no contact with agriculture. Most of them did not speak or read the country's language - Hebrew. To teach these immigrants farming, new approaches to extension were needed.

Volunteers with practical experience were recruited by the settlement authorities. Usually a couple from an established Moshav settlement came to live with the new immigrants in their new village - the husband worked as agricultural advisor, his wife did the same in home economics.

The extension workers were enthusiastic and set a personal example. They worked closely with the farmer and actually shared responsibilities. Their efforts were concentrated on agriculture and they learned about new crops and practices while on the job. This training was compulsory. A close contact was created between extensionists and agricultural research and experimental stations (today extensionists are expected to spend 20-25 per cent of their time on study and the creation of new knowledge). Often targets were set. When new varieties for wheat were introduced, the first target was to reach an average yield of 300 kg per dunam on at least 300 dunam. After operation 300, as it became known, came operation 400 and so on.

Thus, some of the major elements of the Training and Visit approach to extension were present already - constant training and upgrading of extension workers, close contact with farmers in the field (and also with research); and concentration on agricultural development and elimination of unnecessary office work. Another insight from this time was to begin work with the traditional leadership and with opinion leaders. For the first time sociologists joined the extension team.

Extension was also adapted to the cultural patterns of the ethnic sub-cultures. Yemenites, for instance, could read a page upside down, since that was how they studied the Bible in Yemen. Many pupils gathered around one book to learn. They took quickly to reading agricultural pamphlets. Kurds were predominantly illiterate and had a verbal tradition. Therefore, they were approached verbally.

When the new immigrants became better farmers and direct tutoring was no longer needed, the extension system had to be changed. The emphasis shifted from instruction to agricultural development. Because of the specialization in agricultural production, extension workers also had to become specialists.

In the 1950s there were no academic extension workers. In the 1960s an intensive up-grading process began. Extension workers were sent to the Faculty of Agriculture, on part-time release. Today two-thirds of the extension workers hold academic degrees and most new extension workers have university training. When the trend toward specialization in agriculture became stronger, the training of extension workers became more specialized. Now many of them hold a Master's degree in one of the applied agricultural sciences.

In the mid-1960s the Extension Service felt that training in social studies and extension methods was also needed and that the accepted short courses were not enough. A first attempt at setting up an Extension Training Center failed, mainly for institutional reasons. Therefore, the establishment of an Extension Center at the Faculty of Agriculture was carefully planned and finally established in 1979. It offers a one-year diploma course, with an option to combine it with a Master's degree in Agricultural Science which could also be in Extension. Most students take the course part-time. It is too early to evaluate what these extension studies gave (or did not give) to graduates and their employers.

The relatively rapid changes in Israel's agriculture necessitated appropriate organizational adaptations. The first was to organize the special extension service for new farmers and, a decade later, to fuse it with the Ministry of Agriculture's Extension Service, when both farmer groups had reached comparable levels and needs.

Emphasis was placed on independence of extension workers from other functions of the Ministry. It was agreed that extension agents must not fulfil statutory tasks and should not be concerned with production quotas, surveys and so on. Everything was done to gain and maintain farmers' confidence.

Of course, at the same time, intensive planning and organizing of agricultural production continued, but these efforts were made parallel to extension. The positive results strengthened the extensionists rationale of 'agricultural extension only' which was later further developed into the T&V system by Daniel Benor, the former head of the Israeli Extension Service.

In Israel, extension workers were and still are considered to be enthusiastic, reliable professionals. Thus, social approval was the best incentive. It worked well in an organization in which everyone knew everyone else and distances were short. No formal reporting or monitoring system was needed. Extension workers evaluated their own work in respect to present objectives which were revised every year.

Other incentives include opportunities for self-study, as well as participation in field research teams and study tours abroad to keep abreast of developments worldwide. Extension advisors may be appointed as part-time professional referents for new crops, or be made full time heads of departments for major crops. Thus, a relatively young advisor with initiative can become Number One in the country for a new specialized field, if ready for the challenge.

As we have seen, the extension service has given special attention to varying socio-cultural backgrounds of farmers. Advising in a Kibbutz or in a Moshav is not the same thing. At first glance it might seem that extension work with Kibbutz members is much more simple. The advisor, always a specialist in one of the agricultural branches, meets the responsible branch manager in the Kibbutz or the relatively small, closely knit team which runs a given branch. However, managing a branch or the whole farm in the Kibbutz is not considered a 'career position', a term very alien to Kibbutz ideology. Often a branch manager is eventually elected to another economic task, or to a social, political, or public office in the Kibbutz or the Kibbutz movement. Thus, the extension specialist will have to advise a new branch manager every few years, on the average, and continuity is not always guaranteed.

In the Moshav, farm units are much smaller. Continuity is not a problem, but often the educational background of the farmers and their economic strength fall behind that of the Kibbutz. More Moshav farmers have to rely on additional income sources outside of agriculture and this demands a more holistic view of the smallholder's farm - a difficult task for branch specialists.

So far we have considered extension in Kibbutzim and Moshavim. Development was no less dramatic in the Arab sector. It might even be more relevant to extension in developing areas. Let us take, for instance, the case of the Nazareth region. The young and inexperienced Arab extension workers, who set up the service in this region in the early 1960s as part of the national extension service, had to deal with a clientele which was quite different from the Moshavim. Until then Arab farmers had been very conservative, relying only on rainfed agriculture and producing mainly for their own, large families.

The young advisors first had to gain the confidence of farmers in their paternalistic society. They used the knowledge created by applied research and tried out in Moshavim and combined it with respect for the elders in the villages, and with a socio-cultural challenge: i.e. 'What Jewish immigrants can do, you with your experience can do as well.' The division of villages into clans had to be taken into account when organizing meetings and when selecting leader ('contact') farmers. When new crops were introduced, attention was given to socio-cultural patterns, e.g. the availability of a large, extended family for harvesting or the impossibility to let women work after dark, where this was needed (e.g. in the packing of vegetables for exports). On the other side, the same socio-cultural situation enabled the Arab sector to be leading in vegetable production for the inland market and for industry, where hand picking was needed. Once credit facilities were established, the introduction of new techniques such as plastic covers and drip irrigation proved to be no problem.

IMPLICATIONS FOR EXTENSION IN DEVELOPING COUNTRIES

The quick development of a new state, built on agriculture, did not fail to make a strong impression on peer nations in Asia, and especially in Africa, which at the same time were also at the beginning of statehood. Israeli experts, joint projects and courses in Israel became popular. Israel is a multicultural mix and its extension workers learned how to adapt to different cultures.

What, then, were the elements in the agricultural development of Israel which could be adapted to different eco-agricultural and socio-cultural situations in developing countries? The main features were:

- Identification of problems in the field. The best extension technique cannot make up for an inaccurate analysis of problems or for lack of knowledge.
- Empathy with farmers and understanding of their socio-cultural environment, when looking for practical solutions to problems. Farmers' confidence in the advisors will and ability to help are paramount.
- Starting to work with farmers whose advice is sought and accepted by other farmers in the village.
- Regular contacts with the farmers to be served. No waiting 'until the telephone rings'. Most of the time is spent in the field.
- Close contacts with research institutions and participation in adapted field trials.
- Demonstrations, under real farm conditions, on how to implement the advice. The advisor is the first 'to get his hands dirty'.
- Concentration on extension only, while closely coordinating at the top level with planning and with credit, farm supply and marketing organizations - private, cooperative or state, whatever may be available.
- Constant upgrading of staff, at all levels, through training, and encouragement of professional development.

It is not difficult to identify most of these features in the outlines of the Training and Visit extension system. Three other elements which contributed much to the success of agricultural extension in Israel are more difficult to transfer:

- A vision of what applied research and extension can do for rural development and nation-building.
- Dedication to the task, which grows out of this vision.
- Inventiveness, which often leads to improvised solutions until better founded ones can be applied.

Perhaps many development and extension efforts have failed because circumstances favored frustration and nothing was tried to boost vision, dedication and creative problem solving, all of which cannot be imported.

171

III. EMERGING PRIORITIES

Chapter Twelve

DESIGNING AGRICULTURAL EXTENSION FOR
WOMEN FARMERS IN DEVELOPING COUNTRIES

Celia Jean Weidemann
Consultant

BACKGROUND AND OBJECTIVES

Scholars and practitioners in international circles are currently debating the effectiveness of agricultural extension in developing countries. The issues include cost and equity of the Training and Visit (T&V) system promoted by the World Bank and others, adaptability and appropriateness overseas of the U.S. land-grant university extension model; and expanding the role of the private sector.

With increasing frequency a new dimension and a new set of questions have been added to these discussions: how can extension span the gender gap and increase productivity for the significant numbers of women farmers which traditional extension systems have by-passed? Compton (1984) notes that it is usually easier to reach larger, wealthier and often more highly motivated farmers than it is to interact with small-holding, limited resource, low-income farmers. Re-examination and restructuring of extension systems would therefore benefit both women and other small farmers neglected by traditional systems.

The purposes of this chapter are to: (1) describe women's participation in developing country agriculture, (2) analyze their interaction with development projects in general and with two agricultural extension projects in particular, and (3) propose which components of traditional extension models could be modified to identify successful means for reaching women farmers, thereby increasing agricultural productivity.

Agricultural extension refers here to 'an organized, non-formal educational activity, usually supported and/or operated by government, to improve the productivity and welfare of rural people ...' (Swanson and Claar, 1984, p. 15). Others suggest that agricultural extension encompasses individual or group farmer training to spread new or more effective techniques and inputs, assistance to farmers in adapting research results to local conditions, applied research for the development of better farming techniques, and obtain-

175

ing feedback on farmer problems and practices (Berger et al., 1984).

This discourse is confined to agricultural extension. The links between agricultural research and extension are not examined here, nor is the question of whether there is suitable technology to extend to small farmers, particularly in Africa.

WOMEN FARMERS IN DEVELOPING COUNTRIES

Arguments for expanding women's access to agricultural extension must begin with examination of female participation in agriculturally productive activities, and their influence in farm and household decision-making. The focus of this paper is on increasing agricultural efficiency and not equity or distributional aspects of development as such. In earlier years, there was more emphasis on ensuring that development benefited women as much as men. The focus of women in development has now shifted to increasing agricultural productivity and efficiency. This paper also uses the term 'gender analysis', defined as analysis of the interaction of gender variables with development project goals and activities. 'Gender' can sometimes be a more powerful analytical concept than 'women' because it focuses on relationships between males and females.

A comprehensive study by Dixon (1982) used International Labor Organization, Food and Agriculture Organization and national population census data to identify females as a percentage of the agricultural labor force. Results from the 82 countries sampled are presented in Table 12.1.

Dixon's analysis revealed that women constitute 38 per cent of the agricultural labor force in developing countries. Figures are highest for Sub-Saharan Africa (46%) and South and Southeast Asia (45%), with significant but lower participation in North Africa and the Middle East (31%) and Central and South America (18%). These numbers are especially striking considering the incidence of undercounting for women's agricultural labor force participation. There are now efforts to develop more accurate sampling patterns, interviewing procedures and definitions of what comprises economic activity, in order to reduce biases which led to undercounting contributions of unpaid family labor, underestimating seasonality of female labor, self-reporting bias and others.

For example it is estimated that unpaid labor of women in the household, if given full economic value, would add an estimated one-third, or $4,000,000,000,000 to the world's annual economic product (Sivard, 1985).

Furthermore, interviews supported by observation are crucial since cultural norms may predispose some women to under-report their own participation in agriculture according

Table 12.1: Females as percentage of the agriculture labor force according to ILO estimates, FAO Censuses of Agriculture, and revised estimates, 1970

Region	FAO Census		ILO Estimate		Revised Estimate	
	No. of Countries	Mean	No. of Countries	Mean	No. of Countries	Mean
Sub-Saharan Africa	11	47.2	40	36.6	40	45.9
North Africa & Middle East	5	27.0	16	11.1	16	30.7
South & Southeast Asia	5	40.2	19	35.5	19	45.3
Caribbean	2	–	7	26.0	7	39.6

Condensed from Ruth Dixon, 'Women in Agriculture: Counting the Labor Force in Developing Countries', Population and Development Review 8 (September 1982), p. 560.

177

to a study in Peru by Deere and Leon de Leal (1982). In contrast in Nepal, women farmers said they did most of the agricultural work (79 per cent) and more than a third of the decision-making. Female extensionists agreed with the female farmers but the male extensionists thought women did only some of the work and were not involved in decision-making. Male workers were therefore unlikely to perceive female farmers as important recipients of extension information (Shrestha et al., 1984, p. 29).

It is obvious that besides the now well-documented contribution of women to visible agricultural tasks, there is also considerable input by women into agricultural decision-making. For example, other research on eight villages in Nepal concluded that women made 42 per cent of household agricultural decisions and decided jointly with adult males in another 12 per cent of the cases (Acharya and Bennett, 1981). A study of small farmers in Zaire found a high level of consultation between married couples, with 82 per cent of males reporting they would discuss financial problems with their wives (Eele and Newton, 1985). An earlier study in Nigeria found that women planted, applied fertilizers, harvested and marketed and that 89 per cent were also involved in general decision-making processes for farming operations (Patel and Anthonio, 1973).

Several authors have developed schemas for depicting women's agricultural responsibilities. Overholt et al. (1984) described five patterns:

1. Under the system of separate crops men and women are responsible for production, processing and marketing of different crops. Women are traditionally identified with subsistence or food crops and men with cash crops.

2. When there are separate fields, women and men produce the same crops but in different fields.

3. With the separate tasks system, much of the work in a cropping cycle is assigned by gender, such as men preparing the ground and women doing planting and weeding.

4. Under the shared tasks system, males and females undertake the same tasks on the same crops. This is most prevalent during labor bottlenecks, like weeding and harvesting periods.

5. Women-managed farms include two distinct types - de facto systems, where men are away for a period of time and women manage the farm in their absence, and de jure situations, resulting from widowed, divorced, abandoned or never married women. Numbers in this

category are increasing significantly in rural areas. In fact, female-headed households now constitute a third of households in developing countries according to the United Nations figures.

Berger, et al. (1984, pp. 14-15) identified four types of agriculturally active women according to the criteria of: power in decision-making, time spent in farming, and agricultural tasks. Their categories included: (1) women farm owners or managers, who are major decision-makers in agricultural production, devote more of their labor to farming, and are responsible for most of the agricultural operations; (2) women farm partners, who share responsibility for agricultural production with another household member, usually their husband; (3) women farm workers, who are active in agricultural work but make fewer decisions about family farm production; and (4) women agricultural wage laborers, who are often landless. The first three categories are most relevant for agricultural extension providers since women in the latter category do not have significant input into decision-making nor access to land. The authors also pointed out that women farm partners are in fact more likely to have direct contact with agricultural agents then female farm managers, who are the most logical extension clientele of the four categories. Further, they reached the unfortunate conclusion that women have very limited access to agricultural extension programs, and those who do benefit are often taught home economics and other subjects unrelated to their agricultural roles.

Women's understanding of agricultural innovations and principles is limited by their lower educational levels. Research by Jamison and Lau (1982) demonstrated that in the presence of technology, primary education of farmers is directly related to significant increases in agricultural production. In the developing world only 43 per cent of adult women versus 65 per cent of adult men are literate (Population Reference Bureau, 1986, p.6). Gains have been made but the imbalances are great. Significant strides can be made in augmenting food production through a more educated and receptive rural population, including women farmers.

If women cannot rely on extension services, can we assume their male partners who have more access will provide accurate and timely agricultural information to them? Research in Africa indicated such knowledge did not necessarily trickle across to females (Fortmann, 1978 and Spring, 1985). This was especially true when crops or tasks were gender-specific (Fortmann, 1978).

In summary then, we know that women are significantly involved in farming but have little access through organized extension channels to improve their techniques. The catalog of factors to which this has been attributed includes: inability to travel to extension centers, different communication channels from men, lack of land, limited income to purchase necessary inputs or to hire labor and draft power to implement new extension techniques, inconvenient scheduling of demonstration or meeting times and locations, gender bias in extension staffing, lack of improved technology on traditional food crops grown by women since research under colonial governments stressed export crops, lower literacy and educational rates which reduce women's tendency to innovate and make accurate agricultural decisions, and political structures which favor male farmers. Some would argue that inaccessibility to agricultural extension services is a class, not a gender problem, and that focusing on the latter clouds the issue. The evidence in this section supports the notion that while class is certainly a factor, gender deserves separate and critical consideration.

DEVELOPMENT PROJECTS AND WOMEN

There are three ways of structuring projects to deliver resources to women: women-only projects, women's components added to larger projects, and completely integrated projects.

This section summarizes two major synthesizing works (Dixon, 1980 and USAID, 1986) in the growing body of literature on the impact of development projects on women. Both evaluations are of AID-funded projects.

Dixon (1980) reviewed 32 AID-funded projects aimed at increasing production or income, improving welfare and promoting integrated development in food, agriculture, and nutrition education, and community organization. Some of the lessons learned about women's access to project benefits were: (1) women had greater direct access to benefits when planners recognized prevailing gender-related division of labor and built on women's work in enabling women to control their earnings; (2) projects had to fit prevailing cultural norms and allocation of household responsibilities; (3) cultural and legal limits on women's access to project goods and services had to be recognized; and (4) shortages of trained female staff posed major obstacles to the recruitment of more women as project beneficiaries. If women were not identified in project design as beneficiaries, they were likely to remain invisible in project planning, implementation and evaluation.

For project decision-making, female participation was higher when projects were administered by women's sections of governmental, non-governmental or private voluntary organizations than when administration was through general sections of government. Further, when women's programs were affiliated with large male-dominated institutions, decisions on major policy issues tended to be transferred to men in the larger organization.

The second major study examines impact with respect to women for a randomly selected sample of 97 AID-funded projects in five sectors from a total of 416 during the period 1973-85. Those sectors were agriculture, employment, energy, education, and water and sanitation. Forty-three of the 97 projects were in agriculture. Of the 97, ten projects including seven relating to agriculture were then selected for intensive field studies.

The specific objectives of this review were to:

a. assess AID experience in addressing gender concerns through analysis of project experience in these five sectors;
b. identify underlying patterns in the relationship between gender and achievement of project objectives and socio-economic goals; and
c. draw lessons for future programming.

Major conclusions reached were:

1. a strong statistical correlation existed between overall project success and delivering resources to women;
2. women had less access to project resources than men even when both were targeted;
3. 'gender analysis' should extend throughout the life of a project, from design to implementation;
4. recognition of gender differences <u>was more important for agriculture than any other sector</u>.

The study reasoned that gender analysis consisted of ten simple steps:

1. Clarify gender roles and their implications for project strategies.
2. Analyze eligibility to receive project inputs.
3. Define prerequisites for participation in project activities.
4. Examine outreach capabilities of institutions and delivery systems.
5. Assess appropriateness of proposed technical packages.
6. Examine distribution of benefits and its effect on incentives.
7. Consider the reliability of feedback mechanisms.
8. Anticipate likely changes in the roles and status of women.
9. Link changes in the roles and status of women with the expected project impact.
10. Identify needed adaptations.

In practice, standard operating procedures often need to be adapted to accommodate women participants. Adaptations flow best from knowledge gained in baseline gender analysis. For example, in agricultural projects, adaptations might include revised credit requirements, funds earmarked for women's participation, different messages or delivery systems for women farmers, and changes in incentive structures (USAID, 1986; Cloud, 1986; Fortmann, 1986). The lessons from this comprehensive review and the ten steps for gender analysis are derived in part from and are directly applicable to agricultural extension projects.

WOMEN AND AGRICULTURAL EXTENSION

Evaluation of project impact on women has been hampered in general by lack of baseline data-gathering during project inception. Implementation is also constrained since accurate information on agricultural activities by gender is often not available to guide project interventions. Other than the anecdotal evidence and the projects which were evaluated after the fact in the AID study just described, there are few examples of agricultural extension projects which were planned, monitored, and assessed to determine project effectiveness and impact on women. This section describes two AID-assisted extension projects which fall into that category: the Caribbean Agricultural Extension Project (CAEP) and the Women in Agricultural Development Project (WIADP) in Malawi. The outcomes of each project are described below.

The Caribbean Agricultural Extension Project was aimed at building institutional capability in national and regional institutions in participating countries. The detailed project baseline study on women's role in agriculture in the Eastern Caribbean revealed that women constituted as much as 50 per cent of the labor force on several islands but their access to agricultural services was inadequate. The required pre-implementation social analysis also reported women were heavily involved in farming but lacked contact with extension agents. The problem was attributed to lack of gender awareness by agents, existence of gender-differentiated communication networks and scarcity and/or ineffective use of females in the extension services.

In Phase I therefore, the project aimed at increasing effectiveness of national extension services and selected regional institutions and involving women more fully and actively in extension policies and programs. Agents were to be sensitized and delivery systems were then to reach out to part-time farmers and women farm operators. Phase II of the CAEP kept the goal of improving extension and backstopping but the objective of involving women was dropped by AID in favor of a 'farm family' approach. Women in development (WID) components were not mentioned in training activities nor in country-specific programs focusing on activities where women played major roles. Nevertheless, project staff who had been sensitized to gender issues in Phase I did have some continued impact in Phase II.

Although a regular project evaluation failed to address ways in which gender issues were related to project goals, a subsequent evaluation of CAEP revealed the following lessons for future project planning:

1. Extension agents will be more responsive to gender issues if links are clearly drawn to their own work plans and goals. The key intervention point is to relate priority production goals (for example, increased acreage in bananas) to specific target audiences (such as identifying which family members will contribute their labor or decision-making to the expanded effort).

2. Extension communications must be targeted directly to specific audiences who will be implementing the activity. Male and female audiences may require different meeting times, message content, media and perhaps gender of extension agent.

3. Extensionists need to be aware of decision-makers in families and multiple goals which lead to distinct labor contributions and differing propensities to take risks.

4. Extension agents must recognize household income streams may differ for males and females thus influencing appropriate interventions.

5. The 'farm family' approach was seen by the evaluators as gender-blind in comparison to a 'gender analysis' approach.

6. The WID component in this project was not grounded deeply enough in technical agricultural issues. 'Sensitizing' agents to gender issues is a necessary but incomplete project strategy.

7. Institution-building projects may require different strategies and evaluations at much later dates to determine effectiveness of gender-based interventions.

The evaluators recommended that future agricultural extension projects provide for: (1) attention to gender issues and assessment of the impact of extension interventions on women farmers, (2) specific budgetary allocations to support technical assistance, monitoring and testing of innovative interventions and delivery mechanisms, and (3) gender-disaggregated data on participation in extension meetings and agent contact with farmers.

A second AID agricultural extension program designed, monitored and evaluated with special attention to gender issues was the Women in Agricultural Development Project (WIADP) which operated in Malawi's Ministry of Agriculture from 1981-3. This was in fact an 'add-on' component to a farming systems project whose original design overlooked these issues. According to Spring (1985), who served as Director, the project purposes were:

1. Collecting and disseminating research data on women's contribution to small holder agriculture in terms of labor and output;

2. Establishing mechanisms to collect sex-disaggregated data for an adequate data base and to pinpoint whether or not problems were gender related;

3. Enacting several successful action-based projects involving women farmers and extension staff, including farmer-managed demonstrations and trials, credit programs, leadership training, and workshops to retrain staff;

4. Developing workable strategies that implementation teams could use to reach women farmers as client groups;

5. Designing formats to monitor and evaluate participation in extension services by gender;

6. Involving planners in the process of changing policy so that women farmers would be included in development proposals; and

7. Legitimizing the male extension staff's work with female farmers (p.72)

Spring (1985) reported that: (1) women were becoming full-time farmers as men became part-time farmers; (2) women spent as much time on farm work as on domestic activities and did as much farm work as men; (3) labor and cropping patterns varied by locality but women were involved in all aspects of farming; (4) the number of female-headed households in rural Malawi was increasing to approximately one-third, mostly because of male migration; and (5) women were interested in agricultural development services but were handicapped by delivery services. She observed that:

> When women are given the opportunity to receive credit, agricultural training, and inputs, their agricultural performance becomes similar to the better male farmers. Women who head households, both in development project areas and in nonproject areas and who receive inputs

and instruction manage their farms as well as men, obtain similar yields, and practice crop diversity. These women make good use of credit and rarely default on loans (p. 74).

The WIADP staff then analyzed the Malawi National Sample Survey of Agriculture and found that male farmers received more personal visits and advice than did women farmers. Data were then disaggregated into three categories: male household heads, female household heads and wives of household heads. Results indicated that men received more extension services than women. Often wives who were farm partners received more services than female household heads. However, few wives received agricultural information from their husbands.

The Women in Agricultural Development Project then conducted two experiments to determine: (1) whether or not male extension agents could work with women farmers, and (2) whether women farmers could do on-farm experimental trials with precision. The answers to both questions were affirmative. As a result, the WIADP prepared an extension circular, distributed by the Ministry of Agriculture, which gave techniques for male extension staff to contact women farmers and include them in credit, training, demonstrations and visitation programs. Spring concluded that male extension workers can work effectively with women farmers and that new methods and techniques to accomplish this can be devised that are feasible and consider cultural traditions.

A new AID-sponsored agricultural extension project in Zaire promises insights about transferring technology to the predominantly female cassava cultivators in Bandundu Region. The Area Food and Market Development Project was designed using extensive gender-sensitive baseline studies. Project interventions will be monitored and evaluated within this framework. Another interesting feature of the project is its use of private sector voluntary organizations as intermediary institutions for delivering resources to small-holders (Horenstein and Weidemann, 1986).

In summary, the limited evidence to date suggest agricultural extension can be made more effective when issues such as these are considered:

- devising appropriate administrative structures;
- conducting gender analysis to discover constraints, incentives and division of farm and household labor, and using results to modify projects; and
- helping extension agents view the needs of women farmers in relation to the agents' own work plans and production goals.

MODELS FOR EXTENSION DELIVERY

1. Conventional Agricultural Extension Approach
Many Third World systems fall in this category which has as its objectives increasing national agricultural production, farm incomes and the quality of life for rural people. Target groups such as contact or demonstration farmers are often identified to increase the numbers impacted in view of shortages of trained staff. In this model, the agricultural extension system generally operates out of the ministry of agriculture or a sectoral ministry.

2. Training and Visit (T&V) System
Swanson and Claar (1984) contend that this is not a new model but rather an improved version of the conventional government system. The World Bank has introduced this method in some 40 Third World countries. Features of the system are: continuous training and frequent visits by staff occupied solely with agricultural extension, built-in supervision, continuous upgrading of staff, comprehensive monitoring and evaluation and minimal office and paper work. Under this system, Village Extension Workers receive regular training and are supervised in the field by Agricultural Extension Officers who report to Subdivisional Extension Officers. Support is provided by Subject Matter Specialists. Village Extension Workers adhere to strict village visitation schedules, and advise groups of 'contact farmers' on seasonally relevant agricultural techniques. 'Average farmers', selected as contacts, are to share information and serve as examples to other farmers (Baxter and Benor, 1984).

3. University-Organized Agricultural Extension
The United States has the most comprehensive example of this system which is a cooperative effort among federal, state and local governments using the land-grant universities. India and the Philippines have adapted this model which relies on research to identify and solve problems.

4. Commodity Development and Production System
This is a narrower system which seeks to produce and market higher value commodities efficiently and effectively. Links among researchers, input supplies and farmers are generally well-organized. A parastatal body usually controls technology development and transfer, as well as marketing. Quality control is critical and extension agents frequently provide technical advice and inputs simultaneously.

5. Integrated Agricultural Development Programs

These programs are often donor-assisted projects with their own management and technical support systems. They are usually production-oriented and emphasize an integrated approach, often in a specific geographic area. Input supply, credit, extension, marketing and other agri-services are provided.

6. Integrated Rural Development Programs

These participatory rural development schemes blend the community development and the Animation rurale approach of Francophone Africa. Their underlying philosophy is participation by the rural poor in planning, implementing and evaluating programs. Economic and social objectives are promoted along with improved health, nutrition, and basic education. Generalists serve as facilitators to involve the rural poor in program planning, implementation and evaluation while specialists work directly with small farmers to develop, test and demonstrate improved agricultural technology.

7. Farming Systems Research and Extension (FSR&E)

The author has added this seventh category to those of Swanson and Claar. Farming systems is a multi-disciplinary approach which blends social and agricultural production sciences and looks comprehensively at the entire farm and farm family. Farming systems research has the potential for inherent sensitivity to the totality of women's involvement in the home, on the farm and off the farm:

> The integrated demands of the unit of production/reproduction for alternative sources and uses of land, labor, capital, management and equipment in the production system are related. The totality of crops, animals and their by-products for both subsistence use and for market, as well as temporary off-farm employment, are included. FSR involves formal, interdisciplinary problem identification in participation with the farm family. In collaboration with farm families appropriate technology is determined (usually from available technology) and evaluated on their fields under their constraints. FSR implies a two-way flow of knowledge between farm families and researchers (Flora, 1982).

In contrast to Swanson and Claar, Berger et al. (1984) collapsed agricultural extension services into only four distinct institutional models:

1. general, government-sponsored services,
2. crop-specific extension programs,

3.   extension services within integrated rural development projects, and
4.   extension services within women-specific programs (p. 23).

They concluded there is nothing inherent in government-sponsored services to exclude women unless their scope is limited to larger farmers or 'male' crops. Integrated rural development projects have high potential for reaching women farmers if food crops are included in their extension efforts. However, crop-specific programs can by-pass women entirely depending on which crops are targeted. While women-specific programs are most likely to reach women, all too often they have a home economics instead of an agricultural orientation. Thus, the authors noted, it therefore makes more sense to integrate women's concerns into the larger agricultural extension programs, instead of, or in addition to, women-specific efforts.

REDESIGNING EXTENSION SYSTEMS FOR WOMEN FARMERS
(See Table 12.2)

Little systematic analysis has been conducted on the interaction of various extension delivery systems with women farmers. In the following section, the author suggests that for the purpose of looking at gender issues, major extension delivery models logically fall into five categories. The combined schema includes: (1) conventional government-sponsored extension programs, (2) training and visit system, (3) commodity development and production system, (4) integrated agricultural/rural development programs, and (5) farming systems research and extension. The first three types traditionally have few incentives for reaching women farmers and probably require more structural adaptation than do the latter two. Table II describes key features of each model which could be manipulated to test system effectiveness for women farmers. It is followed by more detailed discussion and conclusions. Obviously, these hypotheses for an ideal extension would need testing. Further, while few governments have the resources to revise their systems completely, choices and pay-offs would become evident.

1.   Considerable debate has arisen from the failure of conventional government-sponsored extension programs to provide assistance to women and other small-holders. Suggested improvements include sensitizing male agents and adapting components like content and location of training for women farmers. The discourse on training of extensionists often focuses on gender of agents and how to sensitize predominantly male staff. Recommended delivery modes take into

account women's literacy levels, communication channels and labor demands.

But such efforts are likely to be sporadic and sterile unless planners revise the policies and objectives which underpin their programs and from which staff training and delivery modes are derived. Therefore, stated goals of country development plans and ministries of agriculture must reflect policies of proactive assistance to poor and small farmers including women.

Ideally, as with any sound extension system, government-sponsored systems would be redesigned using available baseline data. In its absence, new data would be gathered on agricultural activities. Agents would be trained to conduct this gender-sensitive research and more importantly to modify extension interventions based on study results and the likelihood of women farmers having lower education and literacy rates. Technological recommendations would address women's crops and consider their labor constraints and incentives. For example, new varieties in their husbands' fields could entail extra labor for weeding, thus taking women from their own plots at critical times. Or water-consuming technologies, such as high grade cattle or water soluble pesticides, may require that women draw extra water and carry it for long distances.

2.	The Training and Visit system now used in at least 40 countries would likely require considerable structural adaptation to reach women farmers effectively. Females would need to be among those selected as contact farmers. Demonstrations and visits by Village Extension Workers would need to include women's plots. Women might be organized into groups to meet with Village Extension Workers. Workers at all levels of the Training and Visit system should have some knowledge of gender analysis for agricultural activities. Subject Matter Specialists would use these data to ensure that feasible production recommendations were available for women farmers.

3.	Commodity development and production extension systems would use gender analysis to identify women's activities in production, processing and marketing systems. High value commodities for which women are or could be responsible would be targeted. These might include milk, cotton, beans or coffee. In fact, commodity programs represent a potentially ideal way to direct resources to women farmers.

Critical inputs and credit under this system would be available to women producers and recommended practices would be analyzed with respect to demands on women's labor. Agents would be trained in modifying interventions for women producers based on gender survey results and women's literacy and numeracy levels.

Table 12.2: Redesigning extension systems for women farmers

Type of System	Policies & Objectives	System Incentives for Reaching Women Farmers
Conventional Government-Sponsored Extension Programs	Would be directed towards improved productivity and welfare of all farmers including women.	Financial, material, training or recognition awards would be given for assisting women farmers.
Training and Visit System (T&V)	Would be directed towards improved productivity and welfare of all farmers including women.	Financial, material, training or recognition awards would be given for assisting women farmers.
Commodity Development and Production System	Would be directed towards high value commodities for which women are, or could be, responsible (e.g. milk, cotton, beans, coffee, etc.).	Financial, material, training or recognition awards would be given for assisting women farmers.
Integrated Agricultural/Rural Development Programs	Would be directed towards improved productivity and well-being of all farmers including women.	Financial, material, training or recognition awards would be given for assisting women farmers.
Farming Systems Research and Extension	Would be directed towards improved productivity and well-being of all farmers including women.	Financial, material, training or recognition awards would be given for assisting women farmers.

Table 12.2: (continued)

Clientele if Redesigned	Approach to Service Delivery	Training of Extension Personnel
Would include female and small farmers as well as larger farmers.	Gender-specific baseline data would be gathered on agricultural activities. Technological packages would address women's crops and consider their labor constraints and incentives. Demonstration would occur in women's fields.	Training would be provided on analyzing needs of women farmers or information given on women's agricultural activities. Training would be given on modifying interventions for women farmers based on gender survey results and women's literacy/numeracy levels.
Female contact farmers would be used.	Women would be organized into groups to meet with Village Extension Workers (VEW). Demonstration would occur in women's fields.	Training would be provided on analyzing needs of women farmers or information given on women's agricultural activities. Subject Matter Specialists (SMSs) would ensure research yielded feasible production recommendations for women farmers.
Commodity producers would include women.	Credit/inputs would be available to women producers. Improved recommended practices would be analyzed for impact on women's labor. Technologies would be developed, tested and available for commodities and processing activities for which women were responsible. Demonstration would occur in women's fields.	Training would be provided on analyzing the production, processing and marketing system to target women's activities. Training would be given on modifying interventions for women producers based on gender survey results and women's literacy/numeracy levels.
Would include female farmers.	Would make program facets: input supply, credit, extension, marketing and other agri-services available to women. Demonstration would occur in women's fields.	Training would be provided on assessing women's agricultural roles and constraints. Training would be given on modifying interventions for women producers based on gender survey results and women's literacy/numeracy levels.
Would include female farmers.	Technology is developed in collaboration with farm family and tested on their fields.	Training would be provided on assessing farm and household roles, constraints, and decision-making. Training would be given on modifying interventions for women producers based on diagnostic surveys, results of trials and women's literacy/numeracy levels.

4. One could argue that integrated agriculture and rural development programs are by their nature somewhat sensitive to women's agricultural roles. Nevertheless Third World development is littered with examples of projects which failed to reach their objectives because input supplies, credit, extension and marketing were not gender-sensitive. Clearly, redesign would include stated policies and objectives about extending services to women farmers as well as gender analysis and agent training in its use.

5. Farming Systems Research and Extension (FSR/E) as a research methodology is a holistic approach to farming and farm enterprises. Entire farm families are viewed as collaborators with multi-disciplinary researchers in testing technology on farmers' fields. Because baseline studies are inherent to FSR/E methodology, gender roles are often taken into account. However, various farming systems approaches have been criticized around gender issues for:

- interviewing primarily male farmers resulting in inaccurate pictures of labor allocation, household constraints and decision-making;
- being overly production-oriented and therefore under-valuing agricultural processing which is often women's work;
- favoring interventions for men's crops and animals as opposed to those of women and;
- using households as basic production/consumption units thereby presenting analytical difficulties when labor patterns, income streams, financial obligations and stakes are separate or different for various household members.

Considerable conceptualization and research is now occurring in farming systems on aspects of intra-household dynamics like labor allocation, income streams and decision-making (see McKee and others in Moock, 1986). Some schools of farming systems research generate various 'recommendation domains' based on economic, social and agro-climatic variables which can be used in extension programming. These domains or other segmented targets derived from the criteria of access to a variety of resources can be used to classify extension clientele (Jiggins, 1983, pp. 70-3). In some cases male and female farmers lacking certain resources would fall into the same domain, or women may form a separate group, based not on their gender but rather access to resources.

Some have argued that fusing the farming systems and the Training and Visit approaches would result in a more perfect extension system. It does appear that FSR/E could instruct other extension models in baseline analysis and intra-household dynamics.

While we have little empirical information about the interaction of women with extension systems, we do have growing evidence about women and development projects in general (Dixon, 1980; Cloud, 1986; Fortmann, 1986 and USAID, 1986). Perhaps it is time to overlay the information we now have about types of administrative structures, supportive policies, and gender analysis onto the variables identified for each of these extension systems. Rigorous research seems to be required to manipulate components within each system, as well as to conduct cross-system comparisons to determine what really works for women farmers. Results of such research would provide governments and donor agencies with criteria and guidelines for making cost-effective choices in designing or modifying extension programs.

For each system, incentive structures and performance criteria, derived from policies and objectives, could be the key to improving effectiveness. Agents need not only training but rewards for assisting women farmers. Depending on the resources available and the system involved, these could take the form of financial, material, training or recognition awards for their efforts and results in this area.

OTHER ISSUES

The studies just described could also be devised to identify conditions under which male agents can be as effective as female agents in providing assistance to women farmers. This question resurfaces routinely. Female agents are the most frequently mentioned means of improving extension systems for Third World women farmers. However, the jury is still out on the importance of gender in structuring effective extension systems for female farmers. Berger et al. note with regard to increasing the numbers of female agents: 'Unfortunately, there is insufficient evidence available to conclude whether this is the most effective method of providing assistance to women farmers' (1984, p.ii).

Women agents, though more likely than men to establish contact with women farmers may themselves have less access to system resources and thus not be able to offer beneficial services to these farmers. Even women agents already employed in agricultural extension organizations may be hampered by types of crops emphasized and characteristics of existing delivery systems. In the recent study of AIDs experience in Women in Development, five of the ten in-depth field studies reviewed extension projects. The evaluators concluded that female farmers' restricted access to extension advice can have a negative impact on efforts to increase production. However, agent gender was not the major factor in reaching female farmers. More importantly, government policies which gave priority to commercial farmers and cash crops grown by men left little incentive for extension workers of either sex to spend time with subsistence farmers. Further, farmers were expected to travel to extension centers which women farmers were not able to do. Extension-workers had no transportation to contact the women in their field and thus they were not reached by the system (USAID, 1986).

Another volume of the AID evaluation reviewed 21 agricultural projects and discovered that having women on design teams did not ensure gender sensitivity but that women on implementation teams had been a key factor in focusing attention on women farmer's needs. However, this was no guarantee that women's issues would be addressed. The report argued for more female USAID agricultural officers and female extension agents in developing countries to increase the likelihood of gender sensitivity (Fortmann, 1986). Yet we do know that male agents with special training can be effective with women farmers (Spring, 1986). In the long term, it is probably advisable from an efficiency as well as an equity point of view to recruit more women into all levels of agricultural extension.

A study of women's participation in the agriculture and home economics institutions which train extension workers in the Third World showed that women account for only 11% of intermediate and only 19% of higher level students in agricultural institutions. Where male-female data are available, the study revealed that 80% of agents are male and 20% are female. Forty-one per cent of the latter are engaged in home economics-related programs. Particularly alarming was the finding that women's representation and opportunities in African agricultural educational institutions were the most limited of any region of the world. This is the continent where women are responsible for the majority of food production and where it is often culturally appropriate for women to teach women. In Africa, women accounted for only 17% of intermediate and higher agricultural enrollment. The figures for Latin American countries were more encouraging at 35%. In Asia, women's participation in agricultural education institutions was almost equal to men's at 47%. While women's opportunity in higher agricultural education in the Third World is generally limited, it has improved since 1970. But the study predicted that women's participation in formal agricultural institutions will deteriorate in the future, relative to men's, unless substantial growth can be assured. Secondary schooling for girls was the one factor studied which was most highly associated with increases in women's participation in agricultural education (Sigman, 1984). Some have therefore recommended that targeted growth rates for women's participation be established by individual institutions (Ashby, 1981; Fortmann, 1986; and Weidemann, 1985). Fortmann argued that training was a variable available to project designers which was easy to manipulate and monitor and that more women should be included in farmer training, extension institutions and degree programs (1986). If, in fact, studies like those proposed in this chapter demonstrate that agent gender is a significant variable, then Sigman's projections must be addressed.

General shortages of extension personnel and lack of transportation are other realities which forestall providing assistance to greater numbers of women farmers. Some have suggested using alternative channels such as women's groups and associations as contact points instead of individual farmers. Others have argued that, because of their vast numbers, the regularity of their contact, and the institutionalized nature of home economics extension in many developing countries, this discipline could be a vital force in Third World development. Home economists do have access to rural women and a grasp of the dynamics of farm households. The question is whether they can be re-tooled to offer advice on agriculture (Weidemann, 1985). Perhaps the initial step should be brief training targeted for specific agricultural production, processing or marketing activities identified in gender analysis.

In Malawi, a new policy has been adopted where home economics agents will spend 75 per cent of their time on agricultural and 25 per cent on home economics activities. This is a step forward but the agents will clearly need retraining. The AID 'Managed Input and Delivery of Agricultural Services' (MIDAS) Project in Ghana had some success with re-focusing the Home Extension Unit (Fortmann, 1986). Others have advocated this approach as well (Jiggins, 1983).

Materials to train extensionists about gender analysis are now becoming available. In addition to the case studies using the Harvard approach to analyze gender roles in development projects (Overholt et al., 1984), there are case studies on intra-household dynamics and farming systems (Feldstein, forthcoming). Another source is a new agro-forestry training manual by Buck (forthcoming).

CONCLUSIONS

We are only now beginning to understand some of the complexities of rural household behavior in the Third World. Beyond the cases cited here and by Ashby (1981) and Berger et al. (1984), we know very little about the interaction of rural women with agricultural extension systems. This chapter reviewed available research and suggested how key variables like policies, objectives, delivery modes, agent gender, training and incentives might be adapted and tested in various extension models to produce both greater pay-offs for women and other small farmers, as well as accelerated Third World development.

REFERENCES

Acharya, M. and Bennett, L. (1981) The Rural Women of Nepal: An Aggregate Analysis and Summary of 8 Village Studies. Vol. II, Part 9 in The Status of Women in Nepal. Kathmandu: Centre for Economic Development and Administration.

Ashby, J. (1981) 'New Models for Agricultural Research and Extension: The Need to Integrate Women', in Barbara Lewis (ed.) Invisible Farmers: Women and the Crisis in Agriculture. Washington, DC: USAID, pp. 144-186.

Baxter, M. and Benor, D. (1984) Training and Visit System. Washington DC: The World Bank.

Berger, M., DeLancey V. and Mellencamp, A. (1984) Bridging the Gender Gap in Agricultural Extension. Washington DC: International Center for Research on Women.

Buck, L. (Forthcoming) Agroforestry Extension Training Source Book. Nairobi, Kenya: CARE.

Cloud, K. (1986) 'Gender Issues in AID's Agricultural Programs: How Efficient Are We?' (Working Paper). Washington, DC: USAID.

Compton, J.L. (1984) 'Extension Program Development' in B.E. Swanson (ed.), Agricultural Extension: A Reference Manual. Rome: Food and Agriculture Organization of the United Nations, pp. 108-119.

Deere, C.D. and Leon de Leal, M. (1982) Women in Andean Agriculture: Peasant Production and Rural Wage Employment in Colombia and Peru. Geneva: International Labor Office.

Dixon, R. (1980) Assessing the Impact of Development Projects on Women. AID Program Evaluation Discussion Paper No.8. Washington, DC: USAID.

—— (1982) 'Women in Agriculture: Counting the Labor Force in Developing Countries', Population and Development Review 8, September, pp. 539-66.

Eele, G.J. and Newton, L. (1985) Small Farmer Survey-Bandundu Region, Zaire. Report of a Consultancy to USAID, Kinshasa, Zaire.

Feldstein, H. (Forthcoming) Intra-Household and Farming Systems Research and Extension Case Studies. University of Florida Farming Systems Support Program and Population Council Case Studies Project. (RFD 1, Box 821, Hancock, N.H. 03449).

Flora, C.B. (1982) Farming Systems Research and the Land Grant System: Transferring Assumptions Overseas, Presented at the 18th Annual Conference of the Association of US University Directors of International

Agricultural Programs, Lincoln, Nebraska.

Fortmann, L. (1978) 'Women and Tanzanian Agriculture Development', Economic Research Bureau Paper, No. 77.4. Dar Es Salaam: University of Dar Es Salaam, Economic Research Bureau.

——(1986) 'A Matter of Focus: Inclusion of Women in USAID Agricultural Development Projects' (Working Paper). Washington DC: USAID.

Horenstein, N. and Weidemann, C.J. (1986) 'Women's Access to Agricultural Extension: Training for Local Organizations'. Washington, DC: International Center for Research on Women.

Jamison, D.T. and Lau, L.J. (1982) Farmer Education and Farm Efficiency. Baltimore: Johns Hopkins University Press.

Jiggins, J. (1983) 'Agricultural Extension and Planning for Rural Women'. Report Prepared for Expert Consultation on Women in Food Production. Rome: Food and Agriculture Organization of the United Nations/ESH.

——(1984) 'Research-Extension-Farmer Linkages: The Implications for Women' (Working Paper No. 3) and 'Farming Systems Research: Do Any of the "FSR" Models Offer a Positive Capacity for Addressing Women's Agricultural Needs?' (Working Paper No. 4). Washington, DC: Consultative Group on International Agricultural Research.

McKee, K. (1986) 'Household Analysis as an Aid to Farming Systems Research: Methodological Issues' in Joyce L. Moock (ed.), Understanding Africa's Rural Households and Farming Systems. Boulder: Westview Press.

Overholt, C., Anderson, M.B., Cloud, J. and Austin, J.W. (1984) Gender Roles in Development Projects: A Case Book. West Hartford, Ct: Kumarian Press.

Patel, A.U. and Anthonio, Q.E.B. (1973) 'Farmers' Wives in Agricultural Development: The Nigerian Case'. Ibadan, Nigeria: Department of Agricultural Economics and Extension, University of Ibadan.

Population Reference Bureau (1986) 'Women in The World: The Women's Decade and Beyond'. Washington, DC: Population Reference Bureau.

Rivera, W.M. (1986) 'Comparative Extension: The CES, TES, T&V and FSR/D' (Occasional Paper No. 1). College Park, MD: Department of Agricultural and Extension Education, University of Maryland.

Schmink, M. and Goddard, P. (1985 Draft) Evaluation of Caribbean Agricultural Extension Project. Washington, DC: Center for Development Information and Evaluation, USAID.

Shrestha, P., Zivetz, L., Sharma, B. and Anderson, S. (1984) Planning Extension for Farm Women. Kathmandu, Nepal: USAID Integrated Cereals Project.

Sigman, V.A. (1984) Women's Participation in Agricultural and Home Economics Education in the Third World (Ph.D. Dissertation). Urbana-Champaign: University of Illinois.

Sivard, R.L. (1985) Women: A World Survey. Washington, DC: World Priorities.

Spring, A. (1985) 'The Women in Agricultural Development Project in Malawi: Making Gender Free Development Work', in Rita S. Gallin and Anita Spring (eds), Women Creating Wealth: Transforming Economic Development. Washington, DC: Association for Women in Development, pp. 71-75.

——(1986) Using Male Research and Extension Personnel to Target Women Farmers. Gainesville; Florida: University of Florida Anthropology Department (Mimeo).

Swanson, B.E. and Claar, J.B. (1984) 'The History and Development of Agricultural Extension' in Swanson, B.E. (ed.) Agricultural Extension: A Reference Manual. Rome: Food and Agriculture Organization.

USAID (1986) AID's Experiences in Women in Development 1973-1985 (Volume I) and Ten Field Studies (Volume II). Washington, DC: USAID Center for Development Information and Evaluation.

Weidemann, C.J. (1985) 'Extension Systems and Modern Farmers in Developing Countries', Agriculture and Human Values, II:1, Winter, pp. 56-9.

Chapter Thirteen

INCENTIVES FOR EFFECTIVE AGRICULTURAL
EXTENSION AT THE FARMER/AGENCY INTERFACE

Jon R. Moris
Overseas Development Institute

INTRODUCTION

This chapter reconceptualizes how organizational contexts can
promote or inhibit agricultural extension, using the public
sector services of East Africa as its example. A key attribute
of an extension organization is the associated incentive
structure, which influences how extension agents and farmers
interact in the field. The term 'agency/farmer interface' is
used to highlight the boundary between external agencies and
local client systems, often taking the form of mutual contacts
between field staff (or the 'access bureaucracy') and local
farmers. Most formal programs related to agricultural
extension depend upon a transfer of ideas, information, and
technology across this boundary. The transfer may be one-
way or two-way, initiated 'from above' or 'from below', and it
may consist of a service offered as well as information given -
but it must occur. The circumstances which encourage an
effective and continuing transfer are the focus of this
analysis.

The primary assumption is that those on either side of
this interchange are likely to perceive certain benefits and
costs associated with their mutual involvement in a common
extension program. These 'proximate incentives' may differ
from the benefits which those at the top assume are being
realized. There are many sociological studies which have
shown that the actual reasons for doing things in large
organizations often depart significantly from formal expectations. When an agency is unable to fulfil its official
mandate, on closer examination it often will be found that the
incentives at the interface with clients are insufficient to
motivate continued cooperation, even when the program itself
may have laudable objectives. Indeed, in the case of East
African extension programs, sometimes neither the contact
staff nor farmers have good reason to cooperate. It will be
shown that a major cause of the poor record of many African
extension services is the continuing discord between ultimate

organizational goals and the proximate incentives which are actually experienced at the agency/farmer interface.

ORGANIZATIONAL OPTIONS

There are several types of field organizations which might have agricultural extension activities as part of their institutional charter. Most African countries display a range of agriculturally oriented institutions: a general 'extension service' with a focus on field crops like maize; commodity-based, public corporations (or 'parastatals') handling export crops like tea or tobacco; project-linked special activities, perhaps with an irrigation or livestock focus; a network of rural training institutions; commercial suppliers of inputs; and perhaps farmers co-operatives or committees. These are the six most frequently encountered types, whose likely organizational attributes are summarized in Table 13.1. The point is simply that all of these justify their existence by claiming to offer some type of agricultural service to farmers. Most are found already established within the matrix of service delivery agencies in the typical African country. However, there will be differences between countries and between regions within one country in the extent to which a particular institutional type is emphasized, and the range of specialized agricultural functions it provides.

The case that particular organizational types have intrinsic superiority for 'doing' agricultural extension can be made in the abstract, but is complicated by an obvious inter-dependency between extension institutions and by the fact that the various crops and livestock which farmers grow may require quite different kinds of services. Also, the farming system itself has implications for the choice of institutional type. Table 13.1 itemizes some of the most frequently cited requisites for organizational success ('clear objectives', etc.), and assigns a tentative weighting for each under the various institutional types. It is obvious from this tabulation that private firms tend to show strong performance in several of the areas which seem to be crucial from a theoretical standpoint. However, these associated weights are not deterministic: a well run cooperative or parastatal could do equally well if these attributes are stressed within its own organizational design.

Under any of our six organizational options for dealing with farmers, the people actually 'on the spot' representing the agency in a rural community face formidable difficulties. Individual farmers in systems relying upon hoe cultivation rarely control more than a few hectares each. The agency must deal with hundreds of smallholders scattered over the landscape, most growing a range of 'cash' and 'food' crops, and few having enough output to constitute an attractive

Table 13.1: Likely organizational attributes

Essential Requisites	Private firm	Land grant college	Ministry of agriculture	Para-statal	Enclave donor project	Farmers cooperative
1. Clear objectives	3	2-3	0-1	1-2	2-3	1-2
2. Find and hold good staff	3	3	1-2	2-3	2	0-1
3. Action-oriented	3	1-2	0-1	2-3	3	1-3
4. High payoff technology	var.0-3	var.0-3	var.0-3	var.0-3	var.0-3	var.0-3
5. Performance managed	3	2	0-1	1-2	2-3	1-2
6. Teamwork at base level	2-3	var.1-3	0-1	var.1-3	var.1-3	2-3
7. Realistic job demands	2-3	2	0-2	2-3	2-3	3
8. Means to coordinate inter-agency matrix	2	0-1	0-1	1-2	0-1	0-1
9. Staff downwardly accountable	2-3	0-1	0-1	0-1	0-1	2-3
10. Rapid information circulation	var.1-3	2-3	1-2	var.1-3	var.1-3	1-2
11. Organizational myth/commitment	3	var.1-3	0-1	var.1-3	var.0-3	1-2
12. Access to tech. expertise	var.0-3	3	var.1-3	2-3	2-3	var.0-2
13. Freedom from political interference	2-3	2-3	var.0-2	var.0-2	2-3	var.0-2

Key: 0 = very poor/weak; 1 = poor/weak; 2 = moderate; 3 = strong/good; var. = variable.
These are merely rough judgmental estimates of the likelihood that a given requisite would be found under typical African conditions.

market to commercial input suppliers. Extension agencies are expected to offer a wide range of specialized services to farmers: recommending planting dates, choosing fertilizers and insecticides, arranging loans, diagnosing diseases, supplying new varieties, demonstrating husbandry, and so forth. And yet the field staff appointed to carry out these demanding tasks are at the bottom of an already impoverished hierarchy. The situation as described for Zambia by a group of external trainers is broadly true of many extension systems in the region [1]: 'Despite the priority granted to rural development in national policy objectives, staff in the field are usually poorly trained, poorly paid, lack transport and welfare services, and feel isolated and neglected.'

It has become widely recognized in recent years that the 'access bureaucracy' in Africa's formal systems for agricultural extension constitutes the weak link, severely limiting the overall productivity which can be obtained from a fairly large investment in the establishment of extension services. Access staff are critical for a number of reasons. First, to the extent that there are resource constraints, these tend to become evident initially at the bottom levels among workers without much influence in the organization. Second, where adaptation of technical recommendations to suit farmers' individual circumstances must occur, it will be contact cadres who must make the diagnosis and adjustment. Third, for many farmers and their households, their impression of the adequacy of public service will depend upon their experiences with an agency's local field staff.

Yet paradoxically, while field staff operate under constraints not encountered at higher levels, their critical function and the degree of difficulty they experience are overlooked by their own agencies. Twenty years ago Robert Chambers argued that contact staff were the 'invisible men' in government bureaucracies; the observation remains valid today. Their situation is strongly affected by the economic stringencies which many African governments have experienced since the late 1970s. Contact staff are posted to remote locations but denied the housing allowances and transport which might make them effective. A larger and larger share of each ministry's budget is absorbed into payment of salaries, so that senior staff supposed to supervise the field units can no longer afford to travel and junior employees are effectively immobilized at their stations. Ultimately, even salaries get neglected and junior staff may be left for several months without payment. At the same time they are berated for being lazy and corrupt.

Farmers, too, have been strongly affected. When prices for the 'official' export crops were high, good husbandry was profitable and individual farmers could expect prompt repayment for crops delivered. Furthermore, farmers had an incentive to cooperate because they received subsidized inputs, and - in many cases - loans which did not need to be repaid if the season was adverse. Extension staff were, then, a source of inputs and financing, both received under generous terms which were subject to amendment if farmers did not actually benefit.

Now all this has changed. Low prices and overvalued currencies interact to make export crops unattractive (in comparison to staples which can be sold locally). The cash flow difficulties being experienced by crop authorities and marketing boards are an incentive for the management to hold back a higher proportion of profits, squeezing the returns to farmers still further and accentuating the decline in export crop production. Then, too, some crop handling agencies have begun to default on payments due to farmers for crops already purchased. The shortages of inputs gives farmers a high incentive to divert fertilizers and insecticides (no longer available in the market place) from 'official' crops to other enterprises, lowering yields still further. Donors have begun to insist loans be repaid, in the very years when many African countries encountered worsening drought. These various tensions reinforce each other, becoming part of an interactive system where there are many reasons to take actions contrary to official extension recommendations. To the extent that contact staff do what is requested of them by their superiors, they will inevitably incur farmers' displeasure and in the process destroy their own effectiveness.

African experience suggests that once an extension system becomes enmired in a negative feedback spiral of this kind, with deteriorating services accelerating a production decline (and vice versa), conventional reform measures become ineffective. African countries have tried most of the usual remedies, or 'privileged solutions': job redesign, professionalization, decentralization, reorganization, devaluation, privatization, etc. In the field of agricultural extension, the two main remedies donors have insisted African countries adopt have been the World Bank's 'Training and Visit' (T&V) system, and farming systems' research (FSR). While each has something to offer (we will review positive contributions later in this chapter) the problems at the agency/farmer interface persist. Thus before we turn to a review of how the T&V and FSR systems have fared in African contexts, it is helpful to look more closely at the contact level. Three aspects are particularly relevant:

1. the adequacy of technical packages,
2. coordinating bureaucracy 'from below', and
3. the problem of untenable working conditions.

INCENTIVES AT THE AGENCY/FARMER INTERFACE

Inadequate Technical Packages

A primary, though not exclusive, output from any agricultural extension system is technical information: recommendations concerning the choice of crops, how and when to plant, fertilizer and insecticide applications, and so forth. There are so many possibilities in a tropical farming system, where many farmers may have between ten and twenty crops under production, that it becomes highly desirable to simplify farm advice to concentrate upon the most critical choices. This is what is meant by the term 'technical packages': a preformulated combination of innovations whose combined effect is predictable. Often the 'package' will include a recommended variety, certain levels of input application, and perhaps plant populations and planting date recommendations. In theory, a well formulated package will almost 'sell itself', since it should result in higher profits than the alternatives. Also, the extension agent's task becomes much easier; instead of mastering all possibilities, field staff can concentrate upon a few proven innovations.

As is so often the case in rural Africa, it has been difficult to put a promising theory into practice. An essential precondition for deriving technical packages is having an adequate base of field-tested agronomic research. At first, the problem was that most technical research was concentrated upon the main export crops. Africa's best known agricultural research stations during the colonial period were linked with particular crop industries, such as Tanganyika's Ukiriguru for cotton research or Kenya's Tea Research Institute at Kericho. This meant that while some crops were studied exhaustively, the alternative enterprises at the farm level remained unexamined. Thus a second precondition (still not met in some countries) is for economic screening of recommendations, so that in addition to yield increases the cost of inputs and husbandry operations are known (using realistic prices, average yields, and likely labor costs). A third pre-condition, only just now becoming recognized because of pressure from various farming systems' research projects, is for the application of whole-farm assessment. This requires that the components in a given package have been weighed from the farmers' perspective, taking into account differing resource availabilities and the potential returns from these same resources if used in a different way (farmers' opportunity costs).

For example, many African crops (e.g. maize, cotton, sorghum) show increased yields if planted early, before or at the beginning of the rainy season. However, these crops are also alternatives to each other. The unqualified recommendation simply to 'plant early' - often seen in various parts of Africa - is meaningless at the whole-farm level. Given that farmers' crops compete for scarce labor and inputs, which are in particularly short supply at planting season, what are required are recommendations which specify the comparative returns under varying levels of risk, input availability, and farm size. The farmers' own screening criteria should be taken into account: the highest profit mix may not be the least risk mix.

Let us illustrate this very important observation. Much of African agriculture is highly seasonal, so that peasant farmers experience a short period when labor demand is very heavy and a long period when little takes place but a family must rely upon food in the store. Where there is a four or five month dry season, the poorer farmers may have nearly exhausted their cash reserves, and the diminished food supply gives insufficient energy for hard physical work while also tempting households to eat their next season's seed. At such times, several considerations become uppermost in farmers' minds: 1) minimizing cash outlay, 2) determining when the rains have actually begun, so as not to risk the loss of seed, 3) shortening the time until there is edible food available, and 4) minimizing labor demands in order to spread effort over a number of crops. It should be noted that none of these are considered in the usual agronomic research program, with its preoccupation on maximizing per hectare yields.

In recent years, agricultural scientists have begun measuring the features of greatest concern to smallholders. The results often confirm the wisdom of 'traditional' practices in contrast to those contained in recommended technical packages. For example, data on the mean labor input required by five alternative options in Kenya show that the Ministry of Agriculture's officially recommended package would require 325 man-days per hectare, in contrast to the farmers' own, requiring 142 man-days (Table 13.2). Again, if one compares the returns per unit of labor at planting time, the Ministry's package yields five shillings per hour (the lowest of the five options), whereas the farmers' own practices yield nearly nineteen shillings per hour (Table 13.3). Much the same conclusion emerges from Alverson's comparison of the returns to 'traditional' versus ministry-recommended practices in Botswana: the cash return to labor hour in the traditional Bangwakgetse system is over three times that obtained from the modern package (1984, p. 5). Even when farmers adopt parts of a recommended technical package, they are likely to do so selectively, and for different reasons than the agricultural researchers may have considered - a situation which Franzel found when looking at the partial adoption of Katumani maize in eastern Kenya (Franzel, 1984). Interviews conducted by the author in the same area indicated a premium on minimizing risks, cash outlays, and labor input. It would appear they continue to display these same preferences today.

Table 13.2: Mean labor input per system[b] (man-hours)[a]

System	Planting	First weeding	Second weeding	Thining	Harvesting	Threshing	Man-hours/ 150 m	Man-days/ha
A	3.58	4.19	2.03	1.00	2.43	1.37	14.60	139
B	9.39	5.09	2.42	–	2.15	2.09	21.14	201
C (Min Ag)	17.15	6.15	3.16	0.59	3.97	3.10	34.10	325
D (farmers)	3.11	3.55	2.59	0.55	3.61	1.50	14.91	142
E	5.48	3.39	3.16	0.48	3.43	1.58	17.52	167

[a]All data from Katumani Dryland Research Station. System C is the one recommended by the Ministry,
[b]and D the farmers' practice.
Source: J.W. Gathee (1982) 'Farming Systems Economics: Fitting Research to Farmers' Conditions'.
In C.L. Keswani and B.J. Ndunguru (eds) Intercropping. Ottawa: International Development
Research Center, p. 139.

Table 13.3: Mean yields and returns to planting labor[a]

System	Yield (kg/ha) Maize	Beans	Yield value (K shs)	Yield value – Seed cost (K shs)	Returns/unit planting labor (K shs/hour)
A	4339	–	4339	4290	17.90
B	–	1723	6031	5710	9.00
C (Min Ag)	3444	768	6132	5800	5.00
D (farmers)	2800	331	3959	3870	18.70
E	3231	371	4530	4380	12.40

Note: maize price, 1 K shs/kg; beans price, 3.50 k shs/kg; maize seed, 3.50 K shs/kg; bean seed, 3.50 K shs/kg. Labor rates are 10 K shs/man-day (8 hr). Land equivalent ratios for C, D and E are 1.24, 0.84 and 0.96, respectively.

[a] Source: J.W. Gathee (1982) 'Farming Systems Economics: Fitting Research to Farmers' Conditions', In C.L. Keswani and B.J. Ndunguru (eds.) Intercropping. Ottawa: International Development Research Center, p. 139.

Consider the dilemma of a frontline extension worker ordered to transmit to farmers recommendations which ignore their own priority concerns. Of course, in a purely commercial system, the farmers would reject the innovation and the sales staff would either make changes or be out of work. In a publicly funded system, however, there tends to be little upwards feedback and faulty recommendations may remain as Ministry policy year after year. To avoid alienating farmers, the contact staff may adopt any of several ploys. They may promise farmers loans and subsidized inputs, partially offsetting farmers' losses. They may alter the technical package, hoping no one at headquarters will notice. They may simply withdraw into a perfunctory display of official duties, winning no converts, but doing little harm. The one thing they cannot afford (unless completely insulated from the farmers' displeasure) is zealously to promote technical packages which they know run counter to farmers' own best interests. Since faulty technical packages are common in Africa, this fact alone explains much seemingly dysfunctional staff behavior.

Coordinating Bureaucracy 'From Below'

Extension services offered by a Ministry of Agriculture tend to be highly bureaucratic in structure and in their mode of field operation. Typical traits include: a steep vertical hierarchy; levels differentiated by entry qualification; a high rate of transfers within any given level; a stress on downwards rather than upwards communication; reliance mainly on public funding received through the central government; and acceptance of a rationale based on performance of 'official duties'. The staff who work in such an organization are likely to think of themselves as government servants rather than as farm advisors, though of course the content of their job assignments will deal with agricultural matters.

The first point to note is that the contradictions inherent between this 'top-down' bureaucratic ideal and farmers' need for 'bottom-up' service will be experienced mainly at the contact level. Those at higher levels in a bureaucratized system have relatively little direct contact with farmers, who are usually encountered in the formalized setting of an 'official visit'. Contact staff are enjoined to serve farmers, but in reality they discover that their most important work relationships are vertical ones to supervisors. The field agent gets rewarded, disciplined, or promoted by bureaucrats located higher in the system, at the divisional or district level and subsequently by others at the provincial or national levels. Few Ministries of Agriculture have devised effective methods for evaluating contact workers' field performance based on farmers' own assessment of services received. By far the commonest evaluation device is the supervisor's confidential report, which in some systems the field worker does not even see. If those at a supervisory level are also short of resources - perhaps without vehicles or access to fuel - they may leave field staff unvisited and unobserved for long periods. It is not surprising, then, if contact staff may come to view their own success as depending upon supervisors' reports rather than the quality of service actually offered to farmers. To understand how field activities can be 'managed' requires an intimate knowledge of how the field bureaucracy is organized, and in particular, how field agents relate to their superiors and vice versa.

Field staff must also deal with other bureaucratic agencies whose activities and clientele overlap their own. For most major crops in Africa, the necessary kinds of external assistance - research, disease and pest diagnosis, improved seed, extension advice, inputs and credit, purchasing arrangements, and perhaps farming training - are vested in several field organizations. This may not have been the case in the first few years after independence, but nowadays there are often from five to ten different agencies whose assistance farmers might need. An initial step is to trace how the different crop services are allocated bureaucratically within the matrix of existing service agencies. The simplest approach is to take each major crop or enterprise - food crops, commercial crops, livestock, fuel, and perhaps even fishing - and then trace sequentially what services farmers require and where these can or cannot be obtained (Moris, 1981, pp. 36-40). Once the inter-agency matrix has been identified for a particular community, one must assess the likelihood that farmers can actually obtain the assistance they need.

One typically finds that some agricultural services are offered according to territory, others by function, and still others by crop (i.e. 'horizontal' versus 'vertical' integration of a given crop industry). It will be rare for all producer services to follow a common pattern, or for all communities to encounter the same array of support agencies. The key observation is that from a field agent's perspective, the local allocation of services and technical functions must be taken as given. Whether or not tea as a crop should be assisted by a national parastatal like Kenya's KTDA is simply not a decision which district-level staff in Kenya make. Instead, at district or lower levels, staff attention should be directed at assessing institutional adequacy within the existing services. What matters are the risks of nonperformance which farmers experience, and gaps where a vital function is not supplied by any agency. Since crop production activities occur in a linear sequence, the output eventually obtained will reflect all constraints impinging upon the production and marketing process. District extension supervisors should try to isolate the limiting factors where a determined effort would exert the greatest leverage on the productivity of the total system.

Prevailing work norms and the deep vertical cleavages between agencies make it difficult for field staff in any one organization to influence the functioning of the overall service delivery matrix. Contact staff are usually relatively junior within their respective agencies. They are not allowed to deal officially with outside organizations, except on the most trivial matters. Instead, they are expected to route requests directed at other agencies through their senior officials at the district, provincial, or even national level before being sent outwards to the other relevant organizations. Where such work norms prevail - and we should note they are common within the civil service of African nations - it becomes almost impossible to achieve effective interorganizational liaison by actions taken through the official system from beneath.

As Holt and Schoorl (1985) indicate, approaches to farming systems' research and extension (FSR/E) have in the past ignored the fact that those involved at the contact level enjoy different degrees of access to power in the larger systems which depend upon their output. The lack of influence exercised by field staff in their bureaucracies is matched by the low prestige and disadvantaged position of the people who do much of the actual fieldwork in African farming. Often field operations are performed by women and temporary workers, while the socially recognized head of household may be away looking for work or involved in other non-farm pursuits. Thus the two sets of people whose co-operation is vital for the effective transfer of extension messages occupy structurally disadvantaged roles within their own social hierarchies. They may even regard themselves as being unable to take independent action on farming matters - a perceived constraint which directly contradicts the assumptions embodied in most models of extension communication.

A useful analytic concept which is pertinent to this situation is provided by Smith, Lethem and Thoolen (1980). They protrayed decision-space available to agency staff as being composed of three nested envelopes. The innermost arena for action concerns factors under the organization's direct control (vehicles, activity timetables, etc.). Outside of this come other factors which can be influenced, but not commanded. And, on the periphery, are relevant but uninfluencible factors (the 'appreciated environment'). Most conventional approaches to management concentrate upon procedures useful in the first sphere, for planning and allocating the organization's own resources. But, as Smith, Lethem and Thoolen argue, this arena is a relatively small one in multi-agency rural development activities where much of what is desired cannot be commanded. (There is a clear contrast here in comparison to, say, construction projects which do control most of the resources needed for achieving agency objectives.) And, from our perspective in this chapter, we note the further restriction that field staff usually have only a small influence within their own agency's program.

But this reality is not reflected in the models either ministries or external analysts hold of the extension situation. Extension staff in Africa are likely to see their role as the giving of orders to farmers, based on the agent's superior mastery of 'modern' farming. Extension supervisors in turn view their own organization as being in a commanding position vis-à-vis other bureaucratic units. Even the externally employed appraisal models make this same assumption, being posited on a deterministic link between 'inputs' and 'outputs'. Investment decisions will be justified by recourse to anticipated rates of return from the initiating agency's field projects. The planning of implementation will be based on time budgets and network diagrams, which presuppose shared information, objectives, and responsibilities. These 'top-down' premises are a widespread feature of rural development proposals, embodying a 'blueprint' rather than 'learning process' approach to local action (Korten, 1980). Thus both the field staff and their external advisors are likely to assume a monolithic power structure. The agency itself will be seen as the prime mover, making investments and giving instructions to other participants (whether these are other institutions or the farmers themselves). Such assumptions may be congruent with incentive structures which apply when an organization is deploying its own resources, but only a relatively small proportion of the necessary actions fall into this sphere.

Instead, the larger share of activities one encounters in an effective agricultural extension program involves outside actors who are not under direct orders. A more adequate conceptualization of the total situation is to see it as being a loosely structured system, containing several semi-autonomous service agencies: Kaplan's review (1982) of health care delivery systems in California suggests that tactics which are effective when dealing with a loose assemblage of agencies differ greatly from what might be appropriate within their internal operations.

Of course, there will usually be ways of circumventing cumbersome 'official channels' in order to achieve a certain degree of organizational coordination across bureaucratic lines. Effective field officers may develop their own modes of unofficial contact, perhaps based on bargaining and the unofficial trading of favors between agencies. They may also lobby on their farmers' behalf to obtain necessary assistance. However, because such influence is informal and contrary to official procedures, it depends heavily on the personal qualities of those involved and is difficult to document or analyze. Alternatively, a given extension worker may be seen by others as enjoying the protection of a 'big man' at headquarters, and so be allowed greater freedom of action than would otherwise be permitted. Paradoxically, it is these informal means of liaison and coordination which can be adversely affected by well-intended efforts to make field cadres more accountable vertically within their own organizations.

In some extension systems (such as within the amateur programs of Francophone Africa or India's earlier village level workers), the contact agent's principal role may be conceptualized as showing farmers how to use specialized services available at a higher level. In many African countries, community development staff within the village are expected to serve this function of linking farmers to the network of available services. The difficulty arises because the various technical agencies have their own field programs and are themselves short of resources. In a top-down, vertically oriented hierarchy they are likely to resist or ignore requests for assistance from outside their own organization. This explains in part why in Africa the experiment of having multi-purpose, village workers has seldom been effective over the longer run.

The superimposition of a bureaucratic structure upon a setting where tasks require moving back and forth between different actors exposes field agents to the dangers of a structured misperception of their performance. Unlike the situation in a primary school or a rural hospital, the contact agent is often located miles away from the supervisor's office. If a field worker concentrates upon visiting farmers, he or she risks not being at station when higher officials happen to visit (a quite unpredictable event for junior staff working in isolated rural stations). If contact staff instead concentrate upon securing political and logistic support from the general administration (the main local source for transport and funds), their efforts will be appreciated by the district government but not by villagers nor by their own ministry. If instead they focus upon attending courses and meetings within their own ministry, they may capture their supervisors' attention but farmers will complain they are never seen in the field.

These underlying tensions are magnified for those who work in remote locations or in administrative systems characterized by a high turnover in staffing assignments. A field worker who anticipates being in a given assignment for only a few months has little incentive to focus on meeting farmers' needs. The stresses are further intensified when the national bureaucracy begins to encounter economic dislocation, curtailing vital recurrent funds and sometimes causing the breakdown of rural transport systems. Then rural extension workers are left in a peculiarly vulnerable situation, promoting technologies which may be no longer economically viable, and being assessed by distant supervisors who cannot afford to carry out field visits. While one can marvel at the resilience of African field bureaucracies under adverse conditions, such circumstances make 'extension work' as normally conceived nearly impossible.

Untenable Working Conditions

In any organization where people work, the resources at their command will have a pronounced effect on their morale and their work productivity. Ever since Herzberg's pioneering studies reported 20 years ago (1966), it has been recognized that work satisfaction is not a unidimensional attribute of an employee's work setting. What one finds instead is that a person's reactions to a job assignment will be a composite of various satisfactory and unsatisfactory aspects - 'satisfiers' versus 'dissatisfiers,' if you will. This is a useful insight to apply in analyzing how field extension assignments are perceived by contact-level staff.

In regard to 'satisfiers' (the positive incentives people expect from a job), one thinks immediately of a range of potential advantages of salaried government employment:

- a good salary, either now or in the future,
- attractive promotion prospects
- visibility of good performance,
- challenging work assignments, and
- security of employment,

On the negative side, the 'dissatisfiers' may include:

- adverse working conditions,
- an unattractive location,
- uncompensated expenses,
- a lack of visible impact,
- low salaries, and
- poor promotional prospects.

While this is by no means a complete listing, it can serve to highlight some of the causes of the chronically low morale in certain Ministry of Agriculture field assignments.

Here it is necessary to interject additional information about the structural aspects of the field extension service in a number of African countries. Very often the bottom cadre, those supposedly in direct touch with farmers, have minimal formal qualifications. This is seldom the Ministry's long-run intention, of course. Officially, the aim is usually to phase out untrained staff, replacing them by certificate trained appointees (those who have completed a two year, post-secondary course). In practice, a substantial proportion of the contact cadre - perhaps from 30 to 60 per cent - will consist of staff originally hired on a 'crash program' or 'temporary' basis.

There are a number of bureaucratic reasons why such staff continue in post despite ministry policies to the contrary. Their salaries are low. They are easier to fire if budgetary shortfalls occur. Sometimes they can be added without central approval, and they are unlikely to be rotated out of the district where they are first employed. If certificate-level salaries are also low, there may be a high turnover of field supervisors so that staff at the bottom are the only ones who really know their territory and who can be counted on to remain in post. Unfortunately, temporary and untrained staff in a technical bureaucracy where educational qualification counts heavily have almost no hope of being promoted or enjoying high future salaries - and they know it.

Similarly, those just above the frontline workers are often also disadvantaged in regard to mobility in the national system, being not quite good enough to continue into more extended professional training. Because the national system of education has continued to expand, the entry level into technical fields rises rapidly in response to increasing supply. After a few years those who did not continue to at least a diploma level will find themselves left behind, no longer possessing sufficient educational background to return for degree-level training. This disability locks them out of meaningful, long-term career advancement. They have become a 'trapped elite,' perhaps resentful that others with only slightly better initial qualifications have continued upwards while they are consigned to work in junior positions for the rest of their careers. Their resentment will be exacerbated in systems where there is an artificially wide gap in salaries between the certificate and diploma levels.

Morale problems are almost inevitable in a system where service staff are structurally disadvantaged within their own organizations but feel separate from and superior to their clients. The great majority of Africa's contact-level extension staff are young men, ex-school leavers who have found their way into ministry employment. A wide socio-economic gulf separates them from their clientele, who may be either men or women but are often 10-20 years older than the agent and perhaps also of another tribe or religious group. Experienced farmers when interviewed are often scornful of the lack of practical knowledge shown by lower-level ministry staff, particularly the younger ones. Of course, there are exceptions. It would be false to represent this as a universal pattern; nonetheless, the point remains that contact staff often feel themselves cut off from their supposed clients.

The fact is that increasingly they are cut off irrespective of inclination by shortages of field transport. We have already noted that a Ministry of Agriculture differs from other agencies in having many of its contact staff located in dispersed field offices and duty stations. Such staff are critically dependent upon access to transport for carrying out their assigned tasks. A challenging work assignment looks very different if the person held responsible does not effectively control the minimum resources required for achieving organizational targets. There have been instances in recent years where extension staff did not receive salaries or were denied bicycle allowances - a degree of economic stringency not seen in most countries since World War II. This explains the clutch of junior staff one will usually encounter travelling along in an official vehicle when field tours can be arranged.

Here the public service nature of a Ministry-operated extension service puts it at a disadvantage when over-optimistic programs are adopted or when budgetary crises arise. Most African countries have sustained the ideal of establishing a uniform network of public services in all districts (e.g. one hospital, one agricultural office, etc.). Unlike the situation under alternative modes of extension, a public service will be expected to duplicate the territorial hierarchy within the general administration. Sometimes, too, the number of districts has increased as large ones are sub-divided into more manageable territorial units. The tendency is to keep staff in post, but economize on housing, transport, and recurrent funds. Also, downgrading of the quality of staff is less visible than closing down a field station. These changes occur incrementally, and are relatively invisible. Often senior ministerial staff are themselves not fully aware of the extent to which their field network has become overstretched.

The main point relative to our discussion is that shortages will appear first in the more remote area and at the lower levels of the hierarchy. Senior staff will usually have at least a few new technical assistance projects coming on line, whose resources can be mobilized to meet pressing demands. Degree-level officers taking charge of larger field stations - say an agricultural institute or an experiment station - can expect transport, subsidized electricity communications, and an opportunity to use ministry-supplied resources to grow their own food. Even when the national economy is nearly bankrupt, they will survive. For contact-level staff posted in remote areas the situation is very different. As already noted, the visibility of field performance declines once supervisors cannot undertake field tours. Junior extension workers may find themselves posted into communities where there is no official housing (commonly provided for teachers and medical staff in most African countries). Often they leave their families behind, 'camping out' in their new work stations. Unless posted near an international border (making smuggling easier), consumer goods, kerosene, and petrol may be scarce or unavailable. Transport is erratic. Schooling or medical services may be absent. The list of unpleasant aspects ('dissatisfiers') lengthens in places subject to high inflation or persistent shortages. Many field staff have been forced quite literally to 'live off the land,' either growing their own food or extracting it by one means or another from local farmers. A deteriorating security situation may cause further disruption.

We conclude that the causes of low morale within the contact cadres are these days multi-faceted and structural in nature. If left uncorrected, they foster a pervading apathy which becomes impervious to the usual kinds of administrative and managerial reform. As long as frontline workers have rational reasons for acting as they do, attempts at increasing control from above will be unproductive and perhaps even counterproductive.

Possibilities for Extension Reform

The analysis of 'frontline' presented here has been biased towards the situation in countries which are experiencing acute economic difficulties but which already have a comparatively large extension service. It is unduly pessimistic with regard to typical working conditions in some places, such as in highland Kenya, Zimbabwe, central Malawi, or Botswana. For other countries, however, there arises a genuine question whether they can afford a typical, Ministry of Agriculture-operated, public extension service. Why even attempt to maintain an elaborate bureaucratic superstructure for agriculture extension if administrative resources are in such short supply?

Whatever one might prefer in theory, the fact that countries like Kenya, Tanzania, Uganda, Malawi and Zambia already have a large, established extension service must be recognized. An administrative service with several thousand employees spread over most areas of the country cannot be shut-down or 'privatized' overnight. In the short run, Ministries of Agriculture will continue to operate research stations, offer technical advice, and plan each district's agricultural development program. They may not fulfil these technical functions in a cost-effective manner, but they will continue to hold this responsibility within the national allocation of public service effort.

If so, the main issue dealt with in this paper is how to improve performance at the field level, where many analysts have suggested the greatest constraints are encountered. What are the prospects that internal reforms might achieve higher output within present levels of resource expenditure? Have farming systems research (FSR) and the 'Training and Visit' (T&V) systems been sufficient to counteract the proximate incentives which encourage dysfunctional organizational behavior within the official extension systems?

FARMING SYSTEMS' RESEARCH

As is by now well known, farming systems research (or FSR) directs its efforts at improving the process of technology generation [2]. From our earlier comments, it should be clear that the FSR focus on whole-farm analysis, on identifying farmers' perceived constraints and varying resources, and on field-testing of recommendations ('adaptive trials' as scientists sometimes term them) is long overdue. Most of the short comings which Belshaw and Hall identified in African agronomic research more than a decade ago (1972) are now being addressed within the various FSR projects which have been implemented since the mid and late 1970s. However, adoption of FSR screening at Ministry research stations has not immediately generated better technical packages (as FSR's proponents had hoped it would). It seems that sometimes adaptive research simply proves the technical superiority of 'traditional' practices which farmers have adhered to despite contrary 'scientific' recommendations. In this sense, FSR has had a greater educational impact on research scientists, in forcing them to weigh consideration of other than mere per-hectare yields, than it has upon African farmers.

FSR also lacked a clear methodology for incorporating its results into the mainline extension system. To be sure, it did require contact-level extension assistance in setting up field trials - an expensive and time-consuming task for which FSR project resources were essential. But the usual Ministry stratagem of making FSR units an adjunct of the existing agronomic research stations left the FSR staff without a means for insuring that farmers gained access to their results. Most Ministries have continued to rely upon out-dated concepts of hierarchical communication, with research stations issuing annual reports to the Director of Agriculture who in turn hands them over to the Extension Division and then, eventually, to the provincial and district levels. By the time such results percolate through the official system they have been modified several times and may bear little resemblance to what FSR staff intended. The strongly bureaucratic nature of most Ministry institutions, then, can completely nullify FSR's own desire to frame 'recommendation domains' based on local differences in farming systems.

Regional offices in support of an FSR approach have been set up serving both East and Southern Africa, and the West African countries. FSR staff have introduced 'diagnostic' field surveys, and have involved a cross-section of Ministry staff as participants. In addition, there have been a series of regional and national training seminars aimed at resolving implementation problems. A major uncertainty at first was the presence of a second approach, the World Bank's 'Training and Visit' system. Senior officials within the extension sections in the various East African countries were preoccupied in adapting the Indian derived T&V system to suit African conditions. Some found the lack of liaison between FSR and T&V confusing, but it is now recognized (within East Africa at least) that the two approaches address different stages in technology diffusion, and can be complementary to each other. The regular training sessions which constitute the core of the T&V system presuppose that field tested technological innovations exist; FSR provides a means to generate such innovations.

While it is too early to determine whether FSR has become institutionalized within the parent ministries, we can identify several potential benefits which could make extension effort more productive. A major advantage of FSR is that it provides the conceptual framework for dealing with variability. Until now, most Ministries of Agriculture attempted to derive uniform recommendations which could be highly misleading in some local environments. A second advantage of FSR is its insistence that a multi-disciplinary perspective is required in technology screening. This provides an opening to consider socio-economic factors which may be highly significant but outside of the strictly agronomic sphere. A third and less tangible benefit of FSR is that it encourages field staff involved in diagnostic surveys to think carefully about the content of field recommendations. Farmers' own observations are taken seriously but not uncritically. FSR's focus upon identifying the limiting factors in a farming system allows Ministry staff to target expensive scientific research on those problems and crops which will have the highest payoff for farmers. All in all, FSR has at least the potential for reorienting bureaucratic efforts towards meeting farmers' perceived needs.

THE T&V SYSTEM

The 'training and visit' system introduced in the late 1970s represents a structured approach to technology diffusion, adapted to the circumstances of an administered extension service. Basically (as outlined by Daniel Benor from his earlier Indian and Turkish experience), it applies classical management principles - clear reporting lines, allocation of work by function, attention to spans of control, regularized training sessions, and a scheduled cycle of field visits - to the extension situation. The advantages claimed for it were that farmers would know when and where to expect visits by an assigned contact agent; contact staff activities could be readily monitored by supervisors; and the extension service would be relieved from other duties like credit or input supply for which it is poorly suited [3]. These were powerful arguments. Given that many African countries had public extension services much like those of the Indian states, though on a smaller scale, the success of 'T&V' in India seemed replicable in Africa.

At first, those of us working in East Africa (and here the author speaks as a former head of the Department of Agricultural Education and Extension in Tanzania's University of Dar es Salaam) were sceptical whether the T&V system was suited to African field conditions. In India, a large investment in agricultural research in the 1960s had created a backlog of available technological innovations. Input supplies had become well organized and accessible to the district and block levels. The Indian bureaucracy has a reputation for efficient administration. As well, most Indian farmers live in densely clustered villages, making routine visits according to a predetermined schedule easy. In all of these key respects, African extension work presents contrary working conditions, raising the question whether an extension methodology aimed at routinizing training and visits was actually appropriate. Without proven technical packages, an attentive extension management, and operational field transport, African extension services would have difficulty duplicating India's 'success story' - or so we felt at the time that the T&V system was first proposed.

Now that the T&V system has been in operation for several years in various countries of East and Southern Africa (notably in Kenya, but also in varying degree within Somalia, Zambia, Malawi, Zimbabwe and Botswana), we can hazard several observations about its performance under African conditions.

First, the T&V model requires the least modification in those countries which most nearly meet the preconditions identified above with respect to the Indian situation. In Africa, the T&V system's greatest impact has probably been achieved in Kenya, where (as in parts of India) the Ministry is densely staffed, input suppliers already exist, and communications are good. As with India, so also in Kenya, two decades of agronomic research on hybrid maize was already in place, providing several technical packages suited to the T&V system's cycle of training seminars. Kenya also possesses an abundance of technically trained specialists who have served as field trainers (or 'subject matter specialists' as they are termed within the T&V system). In lesser degree, the same qualities are found within the extension services of central Malawi, Zimbabwe, and Botswana. In all four countries, there is a marked stress upon administrative efficiency within the public service, an ideology of 'top down' public management which is highly congruent with the premises underlying the T&V system.

Second, the introduction of the T&V system can serve as an occasion for dramatizing the needs of the extension service. Again, using the Kenyan experience, there were already on record several proposals for extension reform suggested from David Leonard's classic analysis of the extension service in western Kenya (Leonard, 1977, pp. 214-7). Yet Leonard's emminently sound recommendations were not of a nature to capture national attention. T&V was introduced into Kenya with the aura of an already proven managerial technology, and it came with external financial support. To a Ministry starved of transport and recurrent funds, it is a tremendous advantage when dealing with the finance ministry to appear on the forefront of international innovation. Thus while the T&V system does require extra transport and training resources beyond those which many African ministries possess, it also provides (so far) the means for generating additional financial support.

Third, an unexpected consequence has been the upwards pressure which adoption of T&V can generate upon a country's research services. Under the usual allocation of functions, agricultural research scientists face relatively little pressure from the field extension service (at least from the smallholders' sector). This situation changes dramatically once there is a regular cycle of training sessions to which agricultural scientists are called as advisors. When technical packages are nonexistent or unsuitable, agricultural scientists soon become aware of this shortcoming. This is not to suggest that they should become 'subject matter specialists', but rather that the T&V system creates an appetite for a continuing input of research results.

Fourth, the T&V system can be modified to suit countries where resource constraints do not permit full implementation according to Benor's initial model. Training sessions can be held once a month, when most field staff come in for their salaries anyway, rather than every two weeks. For contact agents, scheduled visits can be with groups of farmers rather than only with individuals. It may be desirable to limit the time which an individual spends as a 'contact farmer,' so that after three or four years new individuals are brought into the contact network. While these changes undeniably dilute the impact of T&V, they retain its central ideal of coupling regularized training sessions with a reasonable (and verifiable) work load for contact staff.

Fifth, it does appear that Daniel Benor's personal charisma and energy are partly responsible for T&V's success. In some Ministries of Agriculture middle-level staff have failed to plan the training sessions needed to refurbish frontline workers, and have been unwilling to release the transport and financial support which T&V requires. The 'top down' premises of T&V can become counterproductive if those at headquarters use the system mainly to increase control over junior staff in the field. Then none of the causes of poor morale which we have already discussed will be addressed, and the imposition of a rigid and demanding schedule for farm level visits will only increase resentment within the 'frontline' extension workers.

CONCLUSION

The concept of incentive structures outlined in this chapter gives a different perspective to how one assesses the likely outcome from the simultaneous introduction of the FSR and T&V systems. It is immediately apparent that the two approaches are complementary to each other, but address different stages in the technology diffusion process. FSR provides as its output field tested recommendations which will not diffuse unless there is an operational system for conveying them to farmers. The T&V system presumes a continuing input of technical information required by subject matter specialists when planning the cycle of training seminars.

The two approaches are organized to facilitate communication in opposite directions. Basically, FSR is designed to increase upwards feedback from the farm level to researchers. The T&V system aims at strengthening downwards communication of proven results. If the two are being implemented within the same organizational structure, there are bound to be points of friction. FSR can require a great deal of time from bottom-level extension workers, an ancillary assignment which the T&V approach specifically discourages. Most African ministries of agriculture resolved this contradiction by locating FSR in the research division and the T&V system in the extension wing. This minimized bureaucratic conflict, but left unresolved the mechanism for transferring FSR results into the extension pipeline. For such linkages to work, there must be frequent and open two-way communication between the extension and research sections of the ministry, a requirement which is contradicted by the usual bureaucratic structure of public service agencies.

The main difficulty with the T&V system from an incentives perspective is its strong 'top-down' character. It tries to make extension agents more accountable, upwards to their supervisors and downwards to farmers (who can now expect visits on particular days). This 'efficiency' orientation makes sense in settings where other agencies can be expected to do their part, and where the basic working conditions for field staff are satisfactory. If, instead, field agents are denied housing and transport, have inappropriate technologies to recommend, and cannot secure the cooperation of other agencies, the anticipated improvements in extension productivity will not occur. The danger arises because the T&V system is a partial approach dealing with only certain components, but policy makers expect dramatic results. Our analysis in this chapter suggests that in the poorer African countries additional attention to proximate incentives is warranted. A major reason for low productivity in the extension systems as they occur at present is because contact-level staff have multiple incentives to behave in ways which undermine overall organizational goals. Because the T&V model does not address a number of the key ingredients responsible for low morale, its long-run impact may be substantially less than its proponents expect.

Both FSR and the T&V system lack explicit solutions for involving other agencies. Where farmers must deal with an array of producer services, this becomes a significant omission. In the T&V system, the tightening of vertical controls implies that brokerage of functions between agencies will occur at higher levels, or else remain solely the farmers' concern. Perhaps there is room for multi-agency participation in supplying subject matter specialists, though the incentives to encourage such interacting on a regular basis are lacking within the system in its initial form.

On the positive side, analysis of incentives at the farm level suggests the new approaches do represent a significant improvement. Having technical packages which are profitable and relevant will free extension workers from the necessity of disguising or 'sugarcoating' Ministry recommendations. Having a realistic work load in terms of the numbers of farmers to visit is also highly desirable. Frontline workers can definitely benefit from regular training sessions. Benor's insistence that such staff should not be saddled with loan collection or other ancillary duties is welcome. If these positive features can be matched by further experimentation to increase upwards feedback into the extension system, implementation of a combined FSR/T&V program might yet become the breakthrough in extension productivity for which resource starved, Third World extension agencies have been searching.

NOTES

1. Overseas Development Group, University of East Anglia (1981), Agriplan Training System, Trainers Manual, Rome: FAO/AGRIPLAN, p. 5.

2. The main sources on FSR in an African context include Anthony et al. (1979: 116-49), Gilbert et al. (1980), Ruthenberg (1976), and Shaner et al. (1981).

3. For official descriptions of T&V, see Benor et al. (1984) and Benor and Baxter (1984). Independent assessments are available in Howell (1982, 1984), von Blanckenburg (1982), and Moris (1983).

REFERENCES

Alverson, H. (1984) 'The Wisdom of Tradition in the Development of Dry-Land Farming: Botswana,' Human Organization, Vol. 43, No. 1, pp. 1-8.

Anthony, K. et al. (1979) Agricultural Change in Tropical Africa, Ithaca, NY: Cornell University Press.

Belshaw, D.G.R. and Hall, M. (1972) 'The Analysis and Use of Agricultural Experiment Data in Tropical Africa,' East African Journal of Rural Development, Vol. 5, Nos. 1 & 2, pp. 39-72.

Benor, D. et al. (1984) Agricultural Extension: The Training and Visit System. Washington, DC: The World Bank.

Benor, D. and Baxter, M. (1984) Training and Visit Extension. Washington, DC: The World Bank.

von Blanckenburg, P. (1982) 'The Training and Visit System in Agricultural Extension: A Review of First Experiences', Quarterly Journal of International Agriculture, Vol. 21, No. 1, pp. 6-25.

Franzel, S. (1984) 'Modeling Farmers' Decisions in a Farming Systems Research Exercise: The Adoption of an Improved Maize Variety in Kirinyaga District, Kenya,' Human Organization, Vol. 43, No. 3, pp. 199-207.

Gathee, J.W. (1982) 'Farming Systems Economics: Fitting Research to Farmers' Conditions', in Keswani, C.L. and Ndunguru, B. (eds) Intercropping. Ottawa: International Development Research Centre, pp. 136-140.

Gilbert, E.H. et al. (1980) Farming Systems Research: A Critical Appraisal, MSU Rural Development Paper, No. 6. East Lansing, Michigan: Dept of Agricultural Economics, Michigan State University.

Herzberg, F. (1966) Work and the Nature of Man. Cleveland, Ohio: World Publishing Co.

Holt, J.E. and Schoorl, D. (1985) 'Technological Change in Agriculture: The Systems Movement and Power', Agricultural Systems, Vol. 18, No. 2, pp. 69-80.

Howell, J. (1982) 'Managing Agricultural Extension: The T&V System of Agricultural Extension', Agricultural Administration Network, Discussion Paper 8. London: Overseas Development Institute.

Howell, J. (1983) 'Strategy and Practice in the T&V System of Agricultural Extension', Agricultural Administration Network, Discussion Paper 10. London: Overseas Development Institute.

Howell, J. (1984) 'Conditions for the Design and Management of Agricultural Extension', Agricultural Administration Network, Discussion Paper 13. London: Overseas Development Institute, London.

Kaplan, R.E. (1982) 'Intervention in a Loosely Organized System: An Encounter with Non-Being,' Journal of Applied Behavioral Science, Vol. 18, No. 4, pp. 415-32.

Korten, D. (1980) 'Community Organization and Rural Development, a Learning Process Approach', The Public Administration Review, Vol. 40, No. 5, pp. 481-511.

Leonard, D. (1977) Reaching the Peasant Farmer: Organization Theory and Practice in Kenya. Chicago: University of Chicago Press.

Moris, J. (1981) Managing Induced Rural Development. Bloomington, Indiana: International Development Institute.

Moris, J. (1983) 'Reforming Agricultural Extension and Research Services in Africa', Agricultural Administration Network, Discussion Paper 11. London: Overseas Development Institute.

Norman, D. (1980) The Farming Systems Approach: Relevancy for the Small Farmer, MSU Rural Development Paper, No. 5. East Lansing, Michigan: Dept. of Agricultural Economics, Michigan State University.

Overseas Development Group, University of East Anglia (1981) Agriplan Training System, Trainers Manual. Rome: FAO/AGRIPLAN.

Ruthenberg, H. (1976, 2nd ed.) Farming Systems in the Tropics. Oxford: University Press.

Shaner, W.W. et al. (1981) Farming Systems Research and Development, Guidelines for Developing Countries, Boulder, Colorado: Westview Press.

Smith, W.E., Lethem, L. and Thoolen, B. (1980) The Design of Organizations for Rural Development Projects - A Progress Report, World Bank Staff Working Paper, No. 375. Washington, DC: The World Bank.

Chapter Fourteen

INDIA'S AGRICULTURAL EXTENSION DEVELOPMENT AND
THE MOVE TOWARD TOP-LEVEL MANAGEMENT TRAINING*

William M. Rivera
University of Maryland

OVERVIEW

Following a brief introduction on India this article explores
three areas: (1) the previous (since Independence) and
current directions of national agricultural extension efforts;
(2) the near nationwide development of the Training and Visit
(T&V) extension system in India's federated republic of 23
states and its evolving management priorities; and (3) the
recently accelerated move by the national (Union) Government
of India toward promotion of training in agricultural extension
management for senior-level officials. These three overview
sections are followed by an outline of a suggested basic
agricultural extension management curriculum for top-level
management. The curriculum put forward in this fourth
section was originally developed as an outgrowth of a six-
week FAO consultative mission to India for which the author
served as team leader. The final section reviews certain
implications of the current agricultural extension management
training intervention and ends with a number of summary
comments.

INTRODUCTION

India is the second most populous and the seventh largest
country in the world. Its nearly 700 million people represent
over a sixth of the world's population. Some 75 to 80 per cent
of them live in villages (of which there are about 600,000) [1]
and agriculture is the main occupation. India lives in villages

---

*The views and opinions expressed in this chapter do not
necessarily reflect the position or policy of FAO, and no
official endorsement should be inferred.

(Prasad, 1981) and agriculture 'occupies, and will continue to occupy, a crucial place in Indian economy' (Venkatraman, 1986).

The general approach, organizational responsibility, and delivery system of agricultural extension has evolved rapidly and radically in India. Following independence in 1947, the Government of India (GOI) initiated a single-line agricultural extension service which was discontinued with the introduction of the Community Development Program in 1952. Under the Community Development Program, village level workers (VLWs) were employed to provide a wide range of public services (such as health, family planning, etc.) although they were expected to devote some 80 per cent of their time to agriculture.

In the late 1950s the continuing expansion of the population, food shortages, and unfavorable weather conditions highlighted the urgency to accelerate food production. As a result the Intensive Agricultural District Program (IADP) was developed by the GOI; as it quickly increased in area eventually covering some 28 districts, the Ministry of Agriculture's Directorate of Extension (DOE) recognized the need for training extension staff. In the early 1960s, three Extension Education Institutes were created.

Since the early 1960s, several development programs have been launched by the GOI through its Ministry of Agriculture to step up agricultural and allied production, e.g. (a) the Intensive Agricultural Area Program (IAAP) in 1964/5; (b) the High-Yielding Varieties Program (HYVP) in 1966/7; and (c) the Small Farmers Development Agency (SFDA) in 1970/1. Parallel to these development programs, as early as 1966, the DOE (Directorate of Extension) designated short-term staff training courses in agriculture for senior officers of these and other training projects (cf. Prasad, 1981). These courses were organized in association with the state agricultural universities (SAUs), research institutes and colleges.

Thus, before the introduction of the Training and Visit (T&V) system, beginning in 1974, the DOE in the Ministry of Agriculture and Rural Development had for over a decade been developing training courses for different levels of agricultural development staff. With the introduction of the T&V system, extension management training was intensified for grassroots and middle-level staff. More recently, this priority has been extended to include training for senior-level management.

Effective management is a critical factor for success in agricultural extension generally, and this is particularly true for the Training and Visit extension system - the primary system in India. As Baxter (1983) states, 'management has a central role in the Training and Visit (T&V) system of extension and good management is vital to the system's success and impact'.

In numerous developing countries, especially where the T&V system has been initiated, management skills are recognized as a priority for developing, operating and expanding agricultural extension services [2]. Nevertheless, until recently, these skills were focused on the grassroots and middle-level officers engaged in extension work [3] because, as may be imagined, it has been necessary first of all to make the basic system operational.

India's efforts to train senior-level officers involved with extension allocations and direction have been underway since 1980/1. Workshops have been held usually once or twice a year since that time at the national and regional levels regarding the operations of the T&V system. Gradually the need for such training has been accepted as a priority. Indeed, in 1985 a national program was conceived to advance the management competencies of senior-level officials. In this connection an entity known as MANAGE was created in January 1986 by the GOI with assistance from the World Bank. While a fledgling effort on the part of the GOI, this move is notable for several reasons. It represents major leadership on the part of India in this domain [4], both nationally and internationally. It underlines the importance of top officers understanding the function, role, and value of agricultural (production) extension services. Moreover, it implies that while there exist certain general principles of management, these must be tailored to agricultural extension practices, and further that, in India's case, the agricultural extension practices to which management principles must be fitted to apply especially and specifically to the Training and Visit System.

Furthermore, this new national program has resulted in two international assistance efforts, both of which took place in March/April 1986. As already mentioned, the FAO sent a mission to assist the MANAGE with development of a basic curriculum for top officials involved in agricultural extension management [5]. At the same time the World Bank contracted with an Indian national [6] to undertake an assessment of the need for management training in agricultural extension management.

I. AGRICULTURAL EXTENSION DEVELOPMENT IN INDIA

It is important to note at the outset of a discussion of agricultural extension in India that, as is the case in many countries, there are not only various knowledge transfer agencies but indeed most agencies involved in the agricultural development process are concerned with extending knowledge - including civil administrations, commercial networks, special project groups, and co-op federations (all of which provide some sort of 'knowledge transfer' service to producers and market intermediaries).

Moris (Chapter 13) also notes the multi-agency involvement in knowledge transfer to farmers and recommends multi-agency involvement in supplying Subject Matter Specialists (SMSs) systems to train VEWs. Despite the knowledge-transfer matrix that exists, the agricultural production extension services are generally distinguishable in that (1) extension is the sole function; (2) the goal is to assist in the transfer of knowledge and technology for production and production-related purposes; (3) the activities are field-based, as well as media supported; and also (4) the target audience, or clientele, is specifically farmers.

In some cases, however, agricultural extension services are integrated into organizations whose primary purpose may be other than provision of agricultural extension services - such as India's 'Lab to Land' program, or with farm cooperatives, Farming System Research and Extension approaches, etc. In 1983, Denning suggested integrating farming systems research with agricultural extension programs. This idea is being considered regarding development of farm trials within the T&V system in India. Moris herein also discusses this possibility with reference to East Africa. More recently Denning (1985) has focused on integrating agricultural extension programs into Farming Systems Research.

As noted, then, knowledge transfer is also a concern of functional support organizations - such as credit, inputs, and marketing agencies - and serves as a form of general knowledge-transfer extension activity. When thus broadly considered, three forms of organization in agricultural development are concerned with knowledge-transfer services and can be identified as follows:

1. Agricultural (Production-Related) Extension Services - services which undertake knowledge transfer as their sole function, as with agricultural extension systems, such as T&V.

2. Integrated Agricultural Extension Services - services which include agricultural extension as an integrated function along with one or more primary functions, as with certain agricultural research programs, cooperatives, etc.

3. Supportive Functional Services (with Knowledge-Transfer Activities) - services which undertake knowledge transfer activities as a supportive function to their main concerns, e.g. credit, supply and marketing. Thus, we find a growing literature on activities such as 'marketing extension'.

In India agricultural extension efforts are pursued by the Union (national) and state (nationwide) public sectors, commodity extension activities carried out by the private sector, and numerous agricultural and rural development activities undertaken by voluntary non-governmental organizations (NGOs). Figure 14.1 developed by Prasad (1983) delineates three agricultural extension systems (the main extension system, the 'first line' extension system, and the non-governmental extension system), and provides a broad organizational picture of some of the linkages and relationships among these efforts.

The states, however - all of which have agricultural extension services within their State Departments of Agriculture (DOAs) - provide the main field-based agricultural extension services. The services are organized primarily through the Training and Visit (T&V) system which prevails nationwide as the agricultural extension system in India - officially in some 15 of the 23 states in India (at the time of this writing).

Within the public sector alone (the main concern of this discussion) there are several important agricultural and informational extension activities. For this discussion, public sector extension services may best be delineated by review of the structural arrangement of the Union Ministry of Agriculture and Rural Development, keeping in mind that the GOI - like the United States and Canada - is a federated political system and that final authority in particular fields remains with states.

The Union Ministry of Agriculture and Rural Development is organized into four main Departments, viz:

a. The Department of Agriculture (DOA) which includes the Directorate of Extension (DOE) as well as other Directorates such as those of Social Forestry and Animal Husbandry.
b. The Department of Agricultural Research and Education (DARE), whose Secretary is at the same time the Director General of the Indian Council of Agricultural Research (ICAR). DARE's goals and budget are determined by the government-supported but autonomous ICAR coordinating body.
c. The Department of Rural Development (RD).
d. The Department of Food and Civil Supplies (FCS).

The first three Departments - DOA (through the DOE), the DARE and the RD - are involved in agricultural extension work.

In the Department of Agriculture, the Directorate of Extension (DOE) is technically the lead directorate in supporting the professional agricultural extension services being undertaken by all states [7]. The main operations of

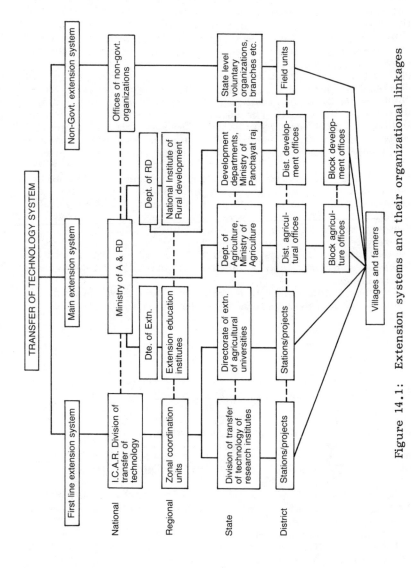

Figure 14.1: Extension systems and their organizational linkages

the DOE have been in the area of training and also supposedly in the development of extension staff (especially middle-level management) [8], but this latter mandate is less apparent in practice. It is the DOE, however, which collaborates most closely with the state Departments of Agriculture (DOAs) as regards training and generally supports states in carrying out field-based extension activities.

The Department of Agricultural Research and Education (DARE) activities are coordinated by the Indian Council on Agricultural Research (ICAR), which is the main coordinating agency at the national level for agricultural research. The ICAR/DARE have initiated a number of special projects, in cooperation with the State Agricultural Universities (SAUs). These projects include agricultural research institutes directed by the ICAR (in association with the SAUs) [9]. All of these activities, however, are research-based, although some experimental extension activities are being pursued - notably the 'Lab to Land' programs which are sponsored by ICAR at certain universities and institutes, to bring farmers in for training in new research developments. Prasad refers to this effort as 'first line' extension work (Prasad in ICRISAT, 1985) [10].

At this juncture it should be mentioned that other directorates of the Ministry of Agriculture (such as Social Forestry and Animal Husbandry) as well as other Ministries (that of Petrochemicals and its fertilizer outreach programs) are concerned with agricultural extension activities [11], although they do not have extension field staff. Another example is that of the Directorate of Marketing and Inspection in India which undertakes extension media programs to explain marketing regulation and make farmers aware of ways to improve the marketing of their products. It supplies extension materials such as documentary films, cinema slides, exhibits and printed literature to the State Marketing Departments, and holds communications workshops for State Marketing staff, who in turn undertake some marketing extension for farmers and also serve as subject matter specialists to State Departments of Agriculture.

There are also a number of other Union Government Ministries and Agencies which carry out knowledge transfer and informational outreach (extension) programs in one form or another. Notable among these are the Ministry of Petrochemicals which extends through its marketing division information to farmers about fertilizers and the GOI-supported Cooperatives Movement which organizes and assists farmers with production, marketing and other services.

Finally, the Department of Rural Development (RD), which is the outgrowth of the Community Development Department, originated in 1952 and started some of the first efforts at extension in India following its independence. Like the Department of Agriculture's DOE, the Department of Rural

231

Development (RD) supports a number of training and development activities, at the national level (such as the NIRD) and at the state level. In addition, it operates several rural development programs (in particular the IRDP, Integrated Rural Development Program) utilizing VLWs (village level workers - not to be confused with VEWs, i.e. the village extension workers employed in the state T&V systems).

But, despite the many levels of knowledge-transfer extension work and the various 'functional support agencies' (Lowdermilk, 1985) involved in knowledge transfer for agricultural development, the overriding leadership and importance to date in what is distinguished here as agricultural (field-based, production-related) extension belongs to the Union Government Directorate of Extension and the state Departments of Agriculture.

II. THE TRAINING AND VISIT SYSTEM AND ITS EVOLVING MANAGEMENT PRIORITIES

At the center of any discussion of agricultural extension and extension management training needs in India is the Training and Visit (T&V) system. A description by Daniel Benor of the T&V extension management system appears in this volume along with other commentaries on the system; and there are numerous other detailed treatments of the system (e.g. Benor and Baxter, 1984). However, some summary comments may be in order.

The T&V system seeks the following:

a. To ensure for a systematically managed organization with a single, direct line of technical support and administrative control.
b. To avoid dilution of efforts due to overburdening of extension workers (VEWs) with multipurpose roles.
c. To limit the number of farms for which an agent (VEW) is responsible.
d. To regularize the training of extension staff, usually on a fortnightly basis, with monthly meetings at the subdivision or zonal level.
e. To develop effective links with research through the fortnightly meeting of VEWs with Subject Matter Specialists (SMS).
f. To improve the status of VEWs through their regular contact with farmers and through their provision to farmers of valuable agricultural information.
g. To reduce the duplication of services by centralizing extension under one administration and avoiding multiple extension schemes which seek to cover particular crops, areas, or techniques.

In brief, the T&V system sets out to create extension services which consist of regular training of agents by subject matter specialists and of regular visits by agents to farmers. It is not so much a new idea - indeed, the basics for the system already existed in India's administrative arrangements under the Community Development Program of the 1950s and 60s - but it is an idea which points to the importance of a strongly hierarchical scheme which regularizes the training of agents and their visits to farmers. Moreover it is a scheme which has been used to carry out a single function: to impact on agricultural production. In essence, it is an extension system which depends on continuous management of the grassroots level and which aims to promote professionalism, a single line of command, concentration of effort, time-bound work, a field and farmer orientation, regular and continuous training, and strong linkages with research.

The organizational pattern generally portrayed in texts on the T&V system (see Figure 14.2) begins with the Department of Agriculture and follows a line of command moving down through the Zone Extension Officer (ZEO), the District Extension Officer (DEO), the Sub-Divisional Extension Officer (SDEO), the Agricultural Extension Officer (AEO), the Village Extension Worker (VEW), and finally to Contact Farmers and Farmers Groups. At two levels (between the DEO and SDEO and between the SDEO and the AEOs and VEWs) appear SMSs (Subject Matter Specialists). These SMSs are employed by the State Department of Agriculture, but are assisted through monthly meetings by faculty from State Agricultural Universities (SAUs). Thus, the T&V system depends on its policy guidelines and administration from the SDAs, and its research connections from the SAUs.

India today has a highly developed network of agricultural extension with a large contingent of staff spread across the nation. According to published, albeit somewhat dated, calculations (Swanson and Rassi, 1981) [12], there were about 71,834 VEWs, some 18,044 AEOs, some 1,895 SMSs, some 5,090 SDEOs and Assistant SDEOs, some 2,000 DAOs and Assistant DAOs, and some 500 state-level Joint Directors and Additional Directors of Agriculture in the states and territories in 1981, at which time T&V was operating in only nine states. Given the expansion of the system to 15 states by April 1986 and that the Swanson-Rassi list does not include APCs (Agricultural Production Commissioners), Secretaries and Additional Secretaries, we may roughly calculate that there are at the middle and top levels some 2,500 to 3,000 officials.

The T&V system with its built-in training and retraining components for functionaries at the field level and for middle-level cadres, provides popular and need-based training programs organized in keeping with the need for timely and local-specific technological information. Grassroots

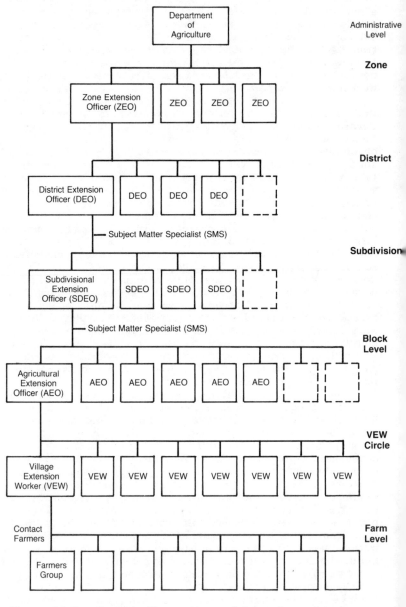

Figure 14.2: Organization pattern of intensive extension
service in one of the states in India

VEWs are mandated to provide periodic technical information to farmers. For middle-level personnel the DOE has developed a number of Extension Education Institutes (EEIs), as already mentioned herein, and these are being utilized for imparting subject matter training as well as extension and management training for the SMS and other middle-level personnel. T&V training for higher level officials in the agricultural extension network has been limited to monthly (zonal) and seasonal workshops on technical matters.

The increased magnitude of the T&V organization in India has affected the extent of management training needs. Cernea [13] notes 'a significant progress in 1976 and 1977 (in India) was that the creation of the T&V system was extended beyond relatively small areas, in order to cover entire states'. Indeed, the T&V system was introduced into India in 1974 as a component of three limited Command Area Development Projects - two in Rajasthan (Chambal and Rajasthan Canal) and one in Madhya Pradesh (covering only two agricultural blocks, viz. Morena and Bhind).

Cernea argues that size is a critical variable for any formal organization, including extension, and the translation from limited command areas to a statewide system entailed a set of organization developments in the structure of T&V as well as some complex problems in its operations. Among these developments and problems, Cernea cites the following six: (1) The hierarchical pyramid in the extension organization became considerably taller. The top management of the state-wide system is less close to the base level, where extension is delivered, than in the case of a command area service. (2) The internal vertical communication channels are stretched out longer. The flow of management information demands more time and is exposed to higher risk of loss or distortion. (3) The basic area unit of a Village Extension Worker has increased (double or triple). The ratio of VEW farmers, which in the command areas was about 1:320, has decreased to 1:600/1000. (4) The high degree of concentration of means, resources, and activities which is typical for a command area development program, cannot be initially replicated state-wide. Therefore, to maintain and improve similar effectiveness, management should be strengthened through better monitoring mechanisms. (5) Whole states are less homogenous than irrigated command areas and the spectrum of agronomic problems to be addressed through extension has become significantly larger. (6) The staff of the extension organization has increased dramatically, from a few tens or hundreds to several thousands, thus enhancing the complexity of monitoring its daily performance (Cernea, 1981, p. 230).

The above comments (1) and (2) are particularly germane to this discussion, although all six have implications regarding extension management. While provocative, however, both comments (1) and (2) appear somewhat exaggerated.

235

T&V's organizational structure did not grow much 'taller' (at most by two levels); it had always been linked hierarchically (vertically) to the state DOA structure up through the Directors of Agriculture; and top management were then and continue to be the budgetary overseers of allocations for agricultural extension. Second, and relatedly, it was the fact of a broadening of the horizontal, not the vertical, internal communication channels that caused these channels to be 'stretched out longer'. Certainly, first-hand knowledge of the grassroots and even the middle-level problems was never a main concern of top management. It simply was not a major concern to those responsible for establishing the T&V in its early years that communication gaps and knowledge lacks existed between the grassroots officers and the middle-level functionaries, not to mention between both of these and top-level officials. As mentioned earlier, it was only in 1980/1 that national and international authorities began to be concerned about strengthening management linkages among these three levels.

At this juncture it is also important to note that the organizational pattern usually cited for T&V refers to the State Department of Agriculture at the top of the administrative hierarchy as a single (unspecified) group of managers. In reality the State Departments of Agriculture incorporate two major arms in their administration: a budgetary arm and an implementation arm. The budget arm includes the APC (Agricultural Production Commissioner), Secretary of Agriculture, and Additional Secretary of Agriculture - who essentially administer financial control of program targets. The implementation arm includes the Director of Agriculture, Additional Director, Joint Commissioners, Assistant Directors, and Deputy Directors; those who translate policy guidelines into program thrusts. The implementation arm may be further divided, with Directors and Additional Directors considered senior-level personnel and Joint, Assistant and Deputy Directors as middle-level personnel. This division has important implications for the design of agricultural extension management curriculum.

III. THE MOVE TOWARD SENIOR-LEVEL MANAGEMENT TRAINING FOR AGRICULTURAL EXTENSION IN INDIA

As mentioned earlier, the primary agricultural extension services in India are provided by the states, not by the central government (GOI), but the Union Ministry of Agriculture has been concerned since its inception with provision of training for agricultural extension development personnel. The GOI's role has included: (a) policy guidance; (b) funding for certain activities; and (c) practical assistance - for instance with training, provision of audiovisuals, and advice on field methods.

During the decade from the mid-1970s to the mid-1980s, the DOE's program for management training increased geometrically as it gained greater resources from the GOI, the World Bank, and the UNDP. In 1980/1, the DOE changed direction slightly, becoming more field-oriented and doubling the number of its training courses - with improvements in the quality of the organization of the courses.

More recently, consequent to its earlier training efforts, the DOE has taken on the responsibility for developing the newly initiated national program to advance senior-level agricultural extension management training in India. The first step in developing this program has been to consider the creation of a National Center for Agricultural Extension Management.

Towards the end of 1985 the DOE had already begun to develop (with the assistance of the World Bank) a management training unit known as MANAGE and decided to house it in the National Institute of Rural Development (NIRD) at Rajendranagar, Hyderabad. Subsequently, in January 1986, this MANAGE unit was designated by the DOE in February 1986 to become the 'National Center for Agricultural Extension Management' [14]. It is the MANAGE unit with which the FAO consultative team worked during March–April 1986 to assist in developing a curriculum - presumably for the newly emerging Center [15].

Two recent developments in India have accelerated the concern with management training. First, as we have already noted, it has become obvious that senior-level officials require continuing education in management skills as they relate to agricultural extension. Second, the Prime Minister has categorically stated that all senior-level officials in all sectors will engage in workshops and training to foster their skills, especially in management. The first observation and the second mandate have brought pressure to analyze and respond to the need for agricultural extension management training.

Simultaneous with the FAO mission, in March–April 1986 the World Bank contracted to undertake a needs assessment which would review the organizational structure, functions, and training needs especially of those officials responsible for state T&V extension services. This assessment (Venkatraman, 1986) [16] concludes that management training is required for all categories and levels of staff employed in extension (grassroots, middle, and top level).

Because of the large number of personnel involved at these different levels, Venkatraman points out that it would be impossible to organize a management training program for all of them at any one institute or center. He therefore proposed a scheme (see Figure 14.3) which categorizes extension management training by level, suggesting different

Grass roots management	VEW AO	State level agricultural training centers, State Agricultural Universities.	Theory and practice stress on skill teaching through exercises (Hands-on course)
	SMS	State Agrl. University, Regional or sub-units of MANAGE as and when they are established.	Theory and practice - stress on skill teaching (Hand-on course) - Include visits to success-ful ORP and other field trials
Middle level management	SDEO/ DEO/ZEO	State level Management Institute (a) State Agricultural University/ other National Institutes/MANAGE	Theory and practice through proper assignment, exercises, demonstrations, case studies (Hands-on courses)
		(b) National/Regional seminars on important topics to be organized by MANAGE	
Top level management	Addl.DA/ DA/Secy./ APC/VC/ Head of other support system.	(A) At MANAGE, Hyderabad	Theory and practice More stress on concepts and models for the present and future use through case studies/ project formulation and analytical exercises.
		(B) Seminars/Workshops organized at regional/national level for short periods with National/International staff. (C) Planned and purposeful study tours to developed/developing countries to study policy planning and program implementation.	

Figure 14.3: Institutions and methods for providing extension management training

institutions as well as distinct approaches and content for training.

Venkatraman mentions specifically Directors of Agriculture and Additional Directors, Secretaries of Agriculture, Agricultural Production Commissioners and Heads of other support systems for training. He sees MANAGE, as the newly emerging 'National Center for Agricultural Extension Management' to be a main source of training - along with seminars and workshops organized at the regional and national levels, as well as planned and purposeful study tours to developed and developing countries to study policy planning and program implementation relating to agricultural extension management. He recommends that workshops and seminars for senior-level personnel stress concepts and models for present and future use, drawing on case studies, project formulations, and analytical exercises.

IV. CORE COURSES FOR SENIOR-LEVEL AGRICULTURAL EXTENSION MANAGEMENT

In designing curriculum for senior-level management involved with agricultural extension the unique character of agricultural extension management must be considered. Each type of extension system presents the student with distinct organizational structures, procedures and mechanisms for implementing the agricultural extension function. It is important to recognize the distinction among systems, for these will affect the managerial theories, principles and practices to be employed in each system.

For example, the Training and Visit system is first of all strictly an agricultural extension system, i.e. its sole purpose is to provide extension services for agricultural production purposes. The U.S. 'Land Grant' Cooperative Extension System is different in that, while its sole purpose is extension, its services cover several main areas: agriculture, home economics, youth development, and community resource development. The Taiwanese 'farm information dissemination system' (Lionberger and Chang, 1981) is distinct in that its programs are supervised and carried out by Provincial Government and Farmers' Association. The management issues in linking research and extension, for instance, will differ for each of these examples of extension systems - even though general management principles remain the same.

Moreover, in order to develop an effective core curriculum for senior-level officials in India responsible for agricultural extension management, two levels of senior officials within the State Departments of Agriculture must be differentiated. First, there are those in the policy and budget allocation branch (Agricultural Production Commissioners, Secretaries of Agriculture, and Additional Secretaries) and secondly there are the officials in the implementation branch (Directors of Agriculture, Additional Directors, Joint Directors, etc.). This division of rank and responsibility requires that core courses be developed and designated as to whether they are applicable to policy branch officials, implementation branch officials, or both.

In addition to differences at the administrative level in the DOAs, it is important to underline the importance of linkages between extension and other systems in the agricultural development process. On examining developments in India, we noted the logic of eventually establishing one or more agricultural development management institute(s) which would provide opportunity to study management as it refers to the various domains of the agricultural development process. Such institution building would presumably be of value also in confronting the political problems involved in improving linkages among the various segments that impact on the agricultural development process. Approaching agricultural extension from a broad perspective – that of the agricultural development process as a whole – and studying more intensely agricultural extension's role in this process would presumably (hopefully) enhance the understanding and interest of secretaries and directors of agricultural extension in its linkage management.

For general purposes management is often defined as the process by which people, technology, job task and other resources are combined and coordinated to achieve organizational objectives effectively (Waldron, 1984). The various functions of management have been categorized by Gulich and Lyndall (1959) as follows under the acronym of POSDCORB: Planning: outlining philosophy, policy, objectives, and resultant things to be accomplished and the techniques for accomplishment. Organizing: establishing structures and systems through which activities are arranged, defined and coordinated in terms of specific objectives. Staffing: the personnel function, selecting and training staff, and maintaining favorable work conditions. Directing: the continuous task of decision-making and embodying decisions in instructions, and serving as the leader of the enterprise. Coordinating: interrelating the various parts of the work of the organization. Reporting: keeping those to whom one is responsible informed as well as keeping the staff and public informed. Budgeting: making financial plans, accounting, managing control of revenue, and keeping costs in line with objectives.

But management involves much more than implementing the POSDCORB functions. Indeed, it requires an awareness of the organization's goals within the framework of social, economic, political and educational forces, as well as a clear understanding of the extension organization itself. In other words, managers must be also good politicians. Furthermore, they are continually confronted with making choices and with allocating, distributing, mobilizing and productively utilizing resources. In addition, their personal style makes an enormous difference in how the functional tasks of the organization will be carried out. In the final analysis, management means leading people, not just managing things.

The FAO team initially organized the curriculum for Agricultural Extension Management into five major categories and developed a total of ten recommended seminar courses. The categories and courses were arranged as follows:

Preliminary Agricultural Extension Management Curriculum

I.   Foundations
     (1) The Foundation Course

II. (Selected) Basic Management Functions
     (2) Planning (A)
     (3) Organizing (B)
     (4) Supervising (B)

III. Communication
     (5) Organizational Communications (A & B)
     (6) Communication Skills for Managers (B)
     (7) Communication Planning & Strategies (A & B)

IV. Staff Development and Training
     (8) Training for Staff Development (A & B)
     (9) Communication Skills for Trainers (B)

V.   Monitoring, Evaluation and Utilization
     (10) Monitoring, Evaluation and Utilization (A & B)

The Foundation Course was intended to introduce the subject of agricultural extension management, to provide a brief overview of each of the courses and thereby assist in guiding officials in their choices of follow-up courses. Each course was designated by the letters 'A' and 'B' (note after each of the above course titles) to refer to whether the course is meant to be directed toward budget branch officials (A) or implementation branch officials (B) or both (A & B), but these designations were meant only to be suggestive.

Even a cursory review of each of the above mentioned courses would require much more space than permissible in this chapter; however, a few words on the Foundation Course may be in order. The Foundation Course covers: (1) the agricultural development process and the extension role therein; (2) the extension function and how it operates, with examples of selected systems and the factors that operate for or against their success; and (3) the management process

(POSDCORB) as it refers to agricultural extension. The Foundation Course covers a broad range of subject matter as organized in the following illustration:

Agricultural	Extension	Management
The Agricultural Development Process	The Extension Function and Major Systems	The Management Process in Agricultural Extension

The Foundation Course, as the other courses, would involve examination of texts and papers as well as field experiences. In particular, the various parts of the agricultural development process, especially its institutional components, are to be examined. The agricultural development process may be variously defined, but the major functional components and linkages in a typical rural system have been cogently illustrated by Axinn and Thorat (1972) as including supply, credit, research, education and extension, production, marketing, and governance. Thus, agricultural extension would be viewed from the beginning as but one component in the agricultural process, interdependent as a system in the process and dependent on economic and other policy supports.

The Foundation Course stresses first that agricultural extension is a function. For agricultural extension systems it is the sole function, while for other institutional systems in the agricultural development process (such as supply, credit, marketing) technology transfer and information dissemination represent only one function which is ancillary but supportive of their primary function (i.e. marketing, supplies, credit).

There are various approaches to agricultural extension which help in understanding particular systems (models) and their purposes. Drawing on the works of Malassis (1976) and Chambers, these approaches include: (a) supervisory (directive); (b) community development; and (c) participant representation. Ray has added (d) hybrid systems, which refer to those that combine elements of two or more of the three major approaches.

The management aspects covered in the Foundation Course draw on the materials from the other courses organized above under the other four rubrics. But any curriculum must ultimately be interpreted and refined by the facilitators who bring guidance to the courses. In this regard, final emphasis should be placed on the importance of carefully choosing and preparing these facilitators. One suggestion is that the initial facilitators be chosen from among the ranks of the senior-level officials themselves.

V. IMPLICATIONS AND SUMMARY COMMENTS

There is considerable commitment to agricultural extension in India and also considerable complexity, especially at the national level, due to the vying among departments within and also outside the Ministry of Agriculture for part of the control over agricultural extension. Aside from the fact that most of the other agriculturally related agencies are also concerned with the transfer of knowledge to farmers and market intermediaries, several bodies within the Union Ministry of Agriculture and Rural Development consider agricultural (production) extension to be specifically within their domain. This presents certain power problems within the Ministry of Agriculture and Rural Development with which the Minister and his staff must continually be concerned.

At the operational level, however, the T&V system is the main field-based, production-oriented agricultural extension system in India, and it is controlled by the individual states. Following on T&V's inception into India in the mid-1970s the most important priority for national training in India was to instill certain management practices at the grassroots and middle-management levels of this system. But as noted herein a steady move since 1980/1 has taken place toward the development of agricultural extension management training for top officials.

In order to foster improved management of agricultural extension, in 1985 the GOI instituted a national program for senior-level training in agricultural extension management. This discussion traces this new initiative by the GOI and outlines a suggested curriculum intended to lead toward accomplishing the objectives of this program [17].

With respect to the move toward top-level management training in agricultural extension, this chapter underlines two important considerations. One is the importance of including in management training the problem of (and solutions to) enhancing linkages among the several functional systems operating within the agricultural development process. The second is to plan a curriculum which accounts for the different needs of officials in the two distinct arms of the state Departments of Agriculture, i.e. for those who manage and allocate the budget and for those concerned more specifically with the implementation of the T&V system. With respect to the policy and budget arm of the DOAs, it is important that these high-ranking officials take greater interest in fostering linkages between research and extension but also among the various agricultural development process systems (e.g. the credit, supply and marketing agencies). The foundation training course proposed herein stresses concern for multi-agency linkage management - to improve coordination and cooperation among the various agricultural development process systems [18].

243

The current intervention to provide systematic management training in agricultural extension management for senior-level officials in India is a priority that has been held in abeyance while the T&V system became operational; it now appears overdue. The experience of India in this domain should be kept in perspective, for it promises to provide insights and guidance to other countries and donor organizations concerned with top-level management of agricultural extension systems. Of particular importance is the decisive step to create a separate agricultural extension development management entity [19]. This discussion argues for the wisdom of this move by India toward senior-level management training in agricultural extension.

To facilitate rural economic growth and social development, training in agricultural extension management appears to be sorely needed - in India as elsewhere. International bodies and bilateral agencies might consider providing assistance to other interested developing countries to establish regional agricultural extension management centers. The MANAGE may serve as a reference point for such development intervention.

More importantly and more obvious, perhaps, is that successful agricultural extension development requires not only managerial expertise but political commitment. Otherwise, a situation may be encountered where the system is successfully managed but politically unappreciated. The efforts by India since its independence to advance agricultural development and to provide for the social equity needs of the rural poor, are remarkable with respect to the former but inadequate for the latter [20]. It can be hoped that these efforts will continue to be expanded and that a balanced approach will take shape in the future.

Finally, of course, the critical point in any discussion of management is whether or not agricultural extension (in this case T&V) becomes more effective at the farm level. Is information accurate and adequate; is it communicated up through the system so that senior officials know which priorities to set? If grass roots and middle-level management problems are not being communicated to top level managers, then senior officials should be encouraged to familiarize themselves with these problems. One way to ensure that these officials become aware of the problems of farmers, VEWs, AEOs, and middle-level extension officers is to engage them in training workshops, seminars, and tours abroad that provide opportunities for them to observe and study basic problems, interests and concerns relating to the development of agricultural extension.

In summary, it appears that priority is finally being placed on agricultural extension management training for top officials. Common sense suggests that it is important for them to become more cognizant of the realities of the farmer and the extension grass roots and middle management officers serving farmers, and of the complex management skills required to make their effort at agricultural production more successful. By its recent action - of creating a national program for agricultural extension management training of senior-level officials - the GOI appears once again to have taken a leadership role in the development of agricultural extension worldwide.

NOTES

1. Figures differ. Prasad (1981) claims there are 630,000 villages, representing 80 per cent of India's total population while Yadava has put the village numbers at 576,000 with about 75 per cent of the population living in them. For the purposes of this article, the figure of 600,000 is an adequate approximation.

2. See the articles on 'Training' and 'Management and System Maintenance' in Cernea, Coulter, and Russell (1983), Agricultural Extension by Training and Visit: The Asian Experience, The World Bank, Washington, DC.

3. Benor, D. and Baxter, M. (1984) Training and Visit Extension. In this basic, practitioner-oriented text, the main chapters treat the roles of the VEWs, AEOs, SDEOs, SMS, and the development of farmers' groups and contact farmers. While senior staff commitment and support for T&V is often mentioned in the World Bank literature on T&V, little exists on training these staff to manage the system more effectively. Particularly as regards India where T&V is so prevalent, this seems a significant omission because the hierarchy has become, as Cernea notes (see reference 4), taller and the communication channels more spread out, and would appear to require both preparatory and periodic upgrading of senior staff in management training relating to extension in general and the T&V system in particular.

4. From 1975 to 1985 the World Bank's main concerns were to establish the T&V extension management system at the grassroots and middle level. Only recently has a major interest taken place to further the management skills of senior-level officials, an interest which has also been assumed by the Union Ministry of Agriculture and Rural Development. This interest is of particular note, as it highlights the value attributed to training in management skills for senior-level development and extension officials. As well, it recognizes the importance of these skills for the success of the T&V system which has advanced in India from a localized ('command area') system to a near nationwide system.

5. This article in no way represents the position or opinions of the Food and Agriculture Organization of the

United Nations. Any omission, errors, or misjudgements are entirely those of the author.

6. Dr. A. Venkatraman, Commissioner and Secretary of Agriculture, Tamil Nadu State.

7. At the time of this writing (April 1986), some 15 states had officially adopted the T&V system, and claims were made by World Bank/NDO (New Delhi Office) that other state would adopt the system before the end of 1986.

8. The DOA/DOE currently supports three middle-level extension and management training institutes - the EEIs, Extension Education Institutes (at Nilokheri in Haryana State, Anand in Gujarat State, and Rajendranagar in Andhra Pradesh). A fourth EEI is planned for Jorhat or Gauhati in Assam.

9. For a fuller treatment of ICAR's activities, see: Prasad, C. (1985), Linkages between Agricultural Research, Education and Extension in India (A country Paper for FAO of the U.N.); New Delhi: Indian Council of Agricultural Research. The list of ICAR activities is taken from this paper, pp. 43-4.

10. See Prasad's comments in ICRISAT (1985), p. 18.

11. There appears to be internal disagreement in the GOI as to whether each technical service should seek to develop its own extension activities in the states, or that the main extension systems in the states should be expanded to include these technical subject concerns in their work.

12. Swanson, B. and Rassi, J. (1981) Directory... The point is not so much to be exact as to indicate the numerical magnitude that must be dealt with at the different levels of T&V organization.

13. Cernea, M. (1981) 'Sociological Dimensions of Extension Organization: The Introduction of the T&V System in India' In: Crouch, B.R. and Chamala, S. (eds), Extension Education and Rural Development, Volume 2, International Experiences in Strategies for Planned Change, Chichester, Australia: Wiley, pp. 221-35, 281. This chapter has also been reprinted by the World Bank in its Reprint Series, Number 196.

14. At the time of this writing, the name of the newly emerging Center is yet to be finalized. In various documents it is titled sometimes as the National Centre for Management of Agricultural Extension and other times as the National Center for Agricultural Extension Management. To date, it is still generally referred to as MANAGE.

15. The FAO consultative mission consisted of a three-person team: the author of this article who served as team leader and curriculum development specialist, Dr. B.L. Coffindaffer, Associate Director of the Maryland Cooperative Extension Service (management specialist), and Dr. S. White, Professor of Communication Arts at Cornell University (communications specialist).

16. Venkatraman, A. (1986) Extension Management Training Needs. New Delhi: World Bank. Dr Venkatraman possesses impeccable credentials, having served in several high-level capacities in the Tamil Nadu State Government and as Vice Chancellor of the State Agricultural University. He is at present the Commissioner and Secretary of Agriculture in that State.

17. However, the question is not merely one of needs but of peers. At the National Consultation of senior-level persons (SDA, SAU, and related research institute directors and faculty) held from April 7 to 11, 1986 at the NIRD by the GOI to review and make recommendations on the agricultural extension management curriculum developed by the FAO consultative team (cf. footnote 5), it was recommended that there be separate categories of senior-level officials. SDA policy-arm officials were designated category (A) and implementation-arm officials as category (B).

18. The former Union Secretary of Agriculture, Shri M. Subramanian, envisioned the creation of an Agricultural Development Management Training Center which would include workshops and seminars on management of all of agriculture's functional systems - governance, credit, supplies, marketing, education, extension, research, and production.

19. For further discussion as to whether training of development administrators should be carried out by the government agency involved, a university, or a separate autonomous body, see: Mathur, H.M. (1983) Training of Development Administrators, U.N. Asian and Pacific Development Center, Kuala Lumpur.

20. To corroborate the 'inadequacy' comment, refer to: Chopra, P. (1983), 'Development and Society: An Overview of the Indian Experience' in Mattis, A. (ed.) A Society for International Development: Prospectus 1984, Duke Press Policy Studies, Durham, NC (Quote p.220):

> India's national movement for independence was engineered and led by the urban middle classes, though the weight of the rural masses was also put behind it by the civil disobedience movement that Mahatma Gandhi created virtually singlehandedly. Therefore, such goals as it had beyond winning political independence did not aim at agrarian revolution or social and economic justice for the rural poor for the sake of such revolution or justice. Its aim for rural India, in any case of a lower priority than its aim for the urban-industrial India of the middle classes, was efficient and productive agriculture as a stable and efficient source of agricultural raw materials for industry and of food for the urban population. Beyond that, its concern for the rural economy was mainly that the village should also become a good market for the products of the town and growers should have

enough incentives to produce more. In fact, there is a close parallel between what the town wants of the village in independent India and what the industrial North wants of the agricultural South on the international plane.

REFERENCES

Axinn, G.H. & Thorat, S.S. (1972) Modernizing World Agriculture: A Comparative Study of Agricultural Extension Education Systems. NY: Praeger.

Baxter, M. (1983) 'Management's Role in Training and Visit Extension: A Comment'. In Cernea, M.M., Coulter, J.K., and Russell, J.F.A., Agricultural Extension by Training and Visit: The Asian Experience. Washington, DC: The World Bank (A World Bank and UNDP Symposium).

Benor, D. and Baxter, M. (1984) Training and Visit Extension. Washington, DC: World Bank.

Cernea, M. (1981) 'Sociological Dimensions of Extension Organization: The Introduction of the T&V System in India'. In Crouch, B.R. and Chamala, S. (eds) Extension Education and Rural Development; Vol. 2: International Experience in Strategies for Planned Change, Chichester: Wiley, p. 221-35, 281.

Denning, G. (1983) 'Integrating Farming Systems Research with Agricultural Extension Programs', International Rice Research Institute.

Denning, G. (1985) 'Integrating Agricultural Extension Programs with Farming Systems Research'. In: Cernea, M.M.; Coulter, J.K.; and Russell, J.F.A., Research-Extension-Farmer: A Two-Way Continuum for Agricultural Development, Washington, DC: The World Bank.

Drucker, P.F. (1966) The Effective Executive. NY: Harper and Row.

Fielder, F. and Chamers, M. (1974) Leadership and Effective Management. Glenville, IL: Scott, Foresman.

Gulick, L. and Lyndall, U. (1959) Papers on the Science of Administration in Extension. Madison, WI: National Agricultural Extension Center for Advanced Study.

Hodgetts, R.M. (1982) Management: Theory, Process and Practice. Chicago, IL: Dryden.

ICRISAT (International Crops Research Institute for the Semi-Arid Tropics) (1985) Training Needs for Dryland Agriculture, with Particular Reference to Deep Vertisol Technology. Patancheru (Andhra Pradesh), India: Author.

India; Ministry of Agriculture; Department of Agriculture and Cooperation (1984) Annual Report, 1983-4. New Delhi: Author.

Jaiswal, N.K., Kolte, N.V., and Arya, H.P.S. (n.d.) Man-

agement of Agricultural Extension: A Study of T&V
System in Rajasthan and Madhya Pradesh.
Rajendranager, Hyderabad: National Institute of Rural
Development.
Lionberger, H.F. and Chang, H.C. (1981) 'Development and
Delivery of Scientific Farm Information: The Taiwan
System as an Organizational Alternative to Land Grant
Universities - US Style'. In: Crouch, B.R. and Chamala,
S. (eds), Extension Education and Rural Development;
Vol. 1, International Experience in Communications and
Innovation, Chichester: Wiley, 155-83.
Lowdermilk, M.K. (1985) 'A System Process for Improving the
Quality of Agricultural Extension', Journal of Extension
Systems, 1 (Dec.), pp. 45-53.
Malassis, L. (1976) The Rural World: Education and Devel-
opment. London: Croom Helm and The UNESCO Press.
Mathur, H.M. (1983) Training of Development Administrators.
UN Kuala Lumpur: Asian and Pacific Development
Center.
Menon, A.G.G. and Bhaskaran, C. (n.d.) 'Management
Systems in Agriculture' (Paper), Kerala Agricultural
University, Mannuthy (Kerala), India.
Mintzberg, H. (1973) The Nature of Managerial Work. NY:
Harper and Row.
Orivel, F. (1981) The Impact of Agricultural Extension
Services: A Review of the Literature. The World Bank,
Washington, DC: Discussion Paper 81-20.
Oxenham and Chambers (1978) Organizing Education and
Training for Rural Development: Problems and Chal-
lenges. Paris: UNESCO International Institute for
Educational Planning.
Prasad, C. (1981) Elements of the Structure and Terminology
of Agricultural Education in India. Paris: UNESCO (See
chapter on 'Agricultural Training for Development').
Prasad, C. (1983) A Study of Agricultural/Rural Extension
Experiences in India. (A Country Paper for FAO of the
UN). New Delhi: Indian Council of Agricultural
Research.
Prasad, C. (1985) Linkages between Agricultural Research
Education and Extension in India. New Delhi: Indian
Council of Agricultural Research.
Ray, H.E., Incorporating Communication Strategies into
Agricultural Development Programs: Guidlines, 3 vols.
Washington, DC: Academy of Educational Development.
Rivera, W.M. (1986) Comparative Extension Systems: The
TES, CES, T&V and FSR/D. College Park, MD:
University of Maryland, Department of Agricultural and
Extension Education, Center for International Extension
Development (Occ. Paper 1).
Rowat, R. (1979) Trained Manpower for Agricultural and
Rural Development. Rome: FAO, Human Resources,

Institutions and Agrarian Reform Division.
Swanson, B. (ed.) (1984) Agricultural Extension: A Reference Manual (2nd Edition). Rome: FAO.
Swanson, B. and Rassi, J. (1981) International Directory of National Extension Systems. Urbana, IL: University of Illinois at Urbana-Champaign, Bureau of Educational Research.
Singhi, P.M., Wadwalkar, S., and Kaur, G. (1982) Management of Agricultural Extension: Training and Visit System in Rajasthan. Ahmedabad: Indian Institute of Management.
Venkatraman, A. (1986) Extension Management Training Needs. New Delhi: World Bank.
Waldron, M.W. (1984) 'Management of Adult and Extension Education', In: Blackburn, D.J. (ed.) Extension Handbook. Guelph, Canada: Guelph University.
World Bank (1981) 'Agricultural Extension in India' (Paper), World Bank, South Asia Projects Department (AGR Technical Note No. 5).

Chapter Fifteen

AGRICULTURAL MANPOWER DEVELOPMENT IN AFRICA*

Wajih D. Maalouf
Food and Agriculture Organization

INTRODUCTION

Food shortage has always been a serious problem in a number of countries in Africa. While famine was declared a widespread disaster over large parts of the region during 1984/5, lack of sufficient subsistence food and malnutrition are not uncommon in the continent, particularly south of the Sahara. The problem became a major catastrophe when a long period of drought (about ten years), struck many African countries, destroying crops, trees and natural pastures and depriving the area of irrigation and drinking water. Millions of livestock vanished and thousands of people died from famine and disease, mainly in Ethiopia.

The international community responded positively to appeals made to help African countries affected by the drought. International organizations, bilateral donors, non-governmental organizations and others participated in a world-wide campaign to alleviate the famine crisis in Africa.

Human resources, as a major element of rural development, is a topic that is being given high priority in assistance programs to affected African countries. The aim of this paper is to present an overview of the status of trained manpower and agricultural training institutions in Africa and to propose a plan for manpower development for the agricultural sector in the region.

---

* The views and opinions expressed in this chapter do not necessarily reflect the position or policy of the Food and Agriculture Organization of the United Nations and no official endorsement should be inferred.

AGRICULTURAL EDUCATION AND TRAINING IN AFRICA

AGRICULTURAL TRAINED MANPOWER IN AFRICA

The disastrous consequences of the African drought could have been greatly reduced if the agricultural sector had had the benefit of adequately trained manpower. The drought calamity could have been better faced at the regional level, especially, if the agricultural sector had been appropriately prepared to deal with such a situation through planning, appropriate exploitation of natural resources, and application of proper practices of agricultural production.

The status of trained manpower in agriculture in Africa must be reviewed with an historical perspective. In colonial times, agriculture was regarded as a practical and second-rate subject compared to medicine, law, science and the arts. This attitude was reflected in the curricula of the universities and in other training opportunities at the time. This led to the present situation where, in most countries, an insufficient number of Africans have college degrees in Agriculture. Those who do have such degrees are generally young and in need of broader experience.

The problem of agricultural trained manpower in Africa has been felt not only by African countries, but by various international, as well as bilateral organizations. In almost every country of Africa, technical and financial assistance from bilateral donors is being provided for strengthening training institutions or training nationals in donor institutions. The United States Agency for International Development (USAID) has provided assistance through American universities for strengthening agricultural faculties in Cameroon, Kenya, Morocco, etc. Similar types of technical assistance are being provided by several European countries, e.g. France to the Comoros Islands with basic training in rural development centers, Belgium to the Extension Training Center in Burundi, the Federal Republic of Germany to training integrated rural development in Cape Verde, etc.

The international organizations, mainly the Food and Agriculture Organization of the United Nations (FAO) and the United Nations Educational, Scientific and Cultural Organization (UNESCO) have large programs for trained agricultural manpower in Africa through technical assistance to institutions and training activities at various levels. FAO is taking a lead role in providing technical assistance focused on institution building, fellowships, and group training activities. To this end, many colleges of agriculture have received technical assistance from the Organization. From 1980-5 more than 1,500 fellowships were awarded to candidates from African countries and 115,000 people participated in group training activities. Additional Africans were also trained at various levels, through multilateral and bilateral assistance.

Despite these efforts, long term planning for agricultural trained manpower development in Africa must be a priority in order to meet the needs of the region and make efficient use of the resources invested in this field. This requires a data base of accurate information on the present situation of manpower and agricultural education and training institutions and their utilization in each country of Africa. There is also a need for valid estimates of the needs of countries in agricultural trained manpower at field, as well as higher and policy levels. During the twelfth FAO Regional Conference African countries requested that FAO conduct a survey to this end. The work was accomplished in 1984 and the results were analyzed and presented to the 13th Regional Conference, held in Harare, Zimbabwe, in July 1984.

THE STATUS OF TRAINED MANPOWER

Information on the present status of manpower and training institutions in Africa, which will be described in this section, is based on surveys of public sector institutions in 47 African countries. The assessment provided information on over 400,000 trained agricultural personnel. Over three quarters of these individuals worked in agriculture (mainly crop production), while the remainder were in forestry, fisheries or livestock.

Trained personnel were classified under three major categories:

1. Professional personnel - those who have completed at least a first university degree at tertiary level of scientific agricultural education (e.g., B.Sc.).
2. Technical personnel - those with intermediate agricultural technical training in secondary or post-secondary agricultural institutions.
3. Vocational or artisanal personnel - those who have had up to two years of vocational education or on-the-job training after primary education.

This assessment was supported and complemented by reports and studies previously carried out by various divisions of the FAO. In summary, results indicated the following:

1. There is, overall, a relative abundance of professional staff. However, numbers are still inadequate, both quantitatively and qualitatively, in the forestry and fisheries sub-sectors, as well as in several specialized fields within agriculture.

2. Although the overall number of professionals in the region seems adequate to meet present requirements, the distribution among countries is unbalanced. In 20

countries, the number of professional personnel represents the minimum required by the year 2000 (taken as a target year to reach sufficiency in agricultural trained manpower). There is currently a shortfall of over 50 per cent in eleven countries.

3. Middle-level technical personnel are in relatively short supply when compared with estimated requirements. Only 14 countries reported (1983) staff adequate to meet the year 2000 minimum requirement. Other countries fell short of this requirement, some of them reporting less than 50 per cent of the estimated requirement.

4. With the exception of Swaziland and Lesotho, who both reported that 25 per cent of their total trained agricultural personnel is female, other countries fell far below this percentage. Agriculturally trained women in the region in the public sector represent only 3.4 per cent of the total work force.

5. Many countries are still in need of expatriates, particularly at the higher levels.

TRAINING INSTITUTIONS

The survey covered 520 agricultural training institutions in 47 countries i.e. about 85 per cent of existing agricultural institutions in Africa. This figure includes 130 institutions at faculty level, with the remainder being intermediate institutions and agricultural high schools. It is important to note the unbalanced distribution of these institutions among countries as illustrated in Table 15.1.

Most African countries have one or more agricultural institutions. Faculty level agricultural programs, with a duration of five years or more, are offered in more than half of the African States. Roughly 200,000 students were enrolled in agricultural institutions, at all levels, in 1983. Women represented 15 per cent of this number. This finding gives encouraging indications that the number of women employed in the agricultural public sector in Africa may increase from its present three per cent average as indicated by the manpower assessment.

In many African countries, expatriate staff still constitute an appreciable percentage of the teaching body at the higher level. National teaching staff in most countries are in need of upgrading and specialization in major agricultural subjects. More than one-quarter of African teachers (excluding Egypt) have qualifications lower than a B.Sc. degree. The present low qualifications of national agricultural teachers and the inadequacy of teaching facilities contributed

Table 15.1:   African agricultural training institutions

Country	High No. of Inst.	Country	Low No. of Inst.
Egypt	86	Botswana	1
Nigeria	65	Congo	1
Zaire	47	Equatorial Guinea	1
Tunisia	46	Mauritania	1
		Swaziland	1

tremendously to the poor quality of agricultural training programs offered in most institutions in the region.

The absence of appropriate agricultural education planning has led to an alarming situation, which is creating unemployment problems in a number of countries. The consequences of such a situation will continue in the future, until proper planning, based on real needs, is applied in this field. Information obtained from FAO studies indicates that in a number of countries, e.g. Algeria, Egypt, Guinea, Libya, Morocco, Nigeria, Sudan, etc., the present trained personnel at the professional level now exceeds the minimum estimated requirements for the year 2000 and enrollment at the faculty level is increasing in these countries. On the other hand, the number of middle-level technicians is relatively lower than the estimated requirements and the enrollment at intermediate institutions is not sufficient to meet the requirements.

The analysis shows that, in general, the present agricultural institutions in Africa possess sufficient capacity to accommodate the required annual number of students. However, this capacity is inadequately distributed between countries. A number of countries, such as Egypt, Nigeria, Tunisia, Ghana and Zaire, have an excess of institutional capacity, while large deficiencies exist in others, e.g. Uganda, Ivory Coast, Ethiopia, Chad and Mozambique. It has also been noted that inadequate coverage is being given in all African agricultural institutions (Egypt excluded) to major subject areas such as horticulture, plant protection, food science, marketing and farm management, rural sociology, agricultural extension, agricultural management and planning. Post-graduate training in major forestry and fisheries subjects is almost non-existent in Africa.

In addition, the quality of agricultural training programs in most African institutions requires substantial improvement, particularly the in-service training of those who hold public posts. Training institutions should be used for this purpose, especially during periods of summer vacation, when facilities are under-utilized.

Finally, African countries should collaborate to effectively utilize institutional capacity. Through this type of collaboration training could be provided under similar conditions and at reduced costs to governments. The cost factor cannot be over-emphasized, considering the financial crisis experienced by all African countries.

RECOMMENDATIONS

In light of the varied situation previously described, it is difficult to propose actions to meet the specific needs of each country. National agricultural training policies and plans must be developed separately for each country, based on in-depth study of national need. However, available information provides background to design a general strategy for the development of agricultural trained manpower in Africa and the improvement of agricultural institutions in the region. This strategy should be formulated with a view to reaching certain targets within a designated period of time.

ELEMENTS AND TARGETS OF THE STRATEGY

The following targets and elements of a strategy for the development of agricultural trained manpower were prepared by FAO and approved by the 13th Regional Conference for Africa:

Elements
- Expansion of training capacity in selected countries, including the construction of physical facilities, training of teachers, and mobilization of the necessary financial resources, for this purpose.

- A sharpened focus on professional training in selected fields, particularly food and cash crop research, project planning and analysis, agricultural extension, veterinary sciences, fisheries and forestry.

- A particular focus on training in managerial skills, both as pre-service post-graduate training and as in-service training of incumbent personnel.

- Improving the quality of training offered at existing training institutions; raising the standard of research and instruction in technical fields at undergraduate and post-graduate levels; and establishing an African network of advanced studies.

- Introducing government-sponsored manpower development programs designed to ensure the best possible utilization of incumbent staff in agriculture.

- Involving women in all types of activity, including pre-service training, within the agricultural sector.

- Increasing regional and inter-country cooperation with a view to making optimum use of existing facilities for agricultural training at all levels.

Targets

It is difficult to establish targets for all of these strategy elements, since some address improvements in quality and therefore are of an intangible nature. Nevertheless, the following strategic targets should be attained by the year 2000:

- All countries in Africa should have the minimum trained manpower in crop production, livestock, fisheries, forestry and other rural development activities, as per the estimates of the FAO Manpower Assessment for Africa.

- All countries should have sufficient capacity for adequately training the required manpower in agriculture, or access to such capacity.

- Female enrollment at agricultural training institutions in Africa should at least double.

- Post-graduate training in the major agricultural fields (at least up to and including M.Sc. or equivalent level) required by 2000 in African countries should be conducted at African institutions, using a well-functioning network of centers of advanced studies.

- There should, in 2000, be no further need for expatriate personnel in senior management capacity.

PROGRAM IDEAS FOR AGRICULTURAL TRAINED MANPOWER DEVELOPMENT IN AFRICA

Any program designed to meet the above-mentioned targets must take time and resource limitations into consideration, as

well as geographical, linguistic and socio-political factors. It
should include short, as well as medium and long-term plans.

SHORT-TERM

In the short-term, actions should be considered in the following areas:

Formulation of Policies and Plans

African governments need assistance to formulate policies and
plans for agricultural trained manpower. Each concerned
government should establish a task force to examine the
present national policies related to agricultural training and
its relationship to research and extension systems. The task
force should study recruitment procedures and career development prospects in depth. It should prepare training plans
oriented toward adequately fulfilling the needs of the
country's agricultural and rural development.

In-service Training

One important activity which needs to be undertaken immediately, is the in-service training of incumbent staff. All professional and technical personnel need systematic in-service
training planned to meet managerial and technical needs. This
type of training should become a regular activity and appropriate funds and facilities should be made available.

Effective Utilization of Existing Manpower

Ineffective utilization of existing staff is also a problem in
Africa. This problem needs immediate treatment through clear
definitions of tasks, improved supervision, provision of
necessary facilities and equipment, allocation of needed
resources and motivation through adequate salaries, promotion
possibilities and improvement of staff living conditions in
rural areas.

MEDIUM AND LONG-TERM

Medium and long-term programs should focus mainly on institution building. This includes the strengthening of existing
institutions as well as the creation of new ones.

Existing Institutions

Existing institutions can be strengthened through the upgrading of teaching staff, the provision of teaching facilities

and equipment, and the improvement of management and supervision. The revision and updating of the curricula can contribute greatly to the quality of teaching programs. A thorough examination of the situation in each major institution is required in order to identify deficiencies and determine appropriate solutions. Some institutions could be expanded to meet these needs.

A number of countries in Africa need to create new institutions at the intermediate and faculty levels. New institutions should be designed to meet the actual needs of the population which they would serve.

Strengthening Major Subject-matter Areas

In many countries, creating new or reinforcing existing departments could well respond to the needs for trained manpower. The following areas require special attention in most African countries: program planning, management, marketing, crop protection, agricultural education and extension, etc. In all types of institutions emphasis should be placed on management as a major component of development programs.

Centers of Advanced Studies

African professional staff are in need of advanced training in specific agricultural subjects. At present, this type of training is only available in specialized institutions in Europe and America. It is the hope of the governments of the region to create possibilities for such training within Africa. The program for agricultural trained manpower should include a substantial component to assist national institutions in developing specialized areas of expertise at an advanced level. This objective could be attained through the development of a network of centers of advanced agricultural studies. The center should be located at a national institution possessing adequate facilities to accept trainees from other African countries, with national governments facilitating enrollment of foreign trainees from the region.

Degree Accreditation

In order to facilitate the exchange of trainees among African countries and encourage technical cooperation among these countries in the field of agricultural training, it is important to develop a system of degree accreditation among institutions carrying out training programs at similar levels. The present dissimilarities in teaching programs and awarded degrees and diplomas create problems in the evaluation of candidates for enrollment at institutions.

Fellowships

If the program ideas mentioned above are put to immediate execution, it will still be a number of years before they produce an appreciable impact on the development of agricultural trained manpower in most African countries. Meanwhile, upgrading of the teaching staff has to be implemented at present in higher education institutions. A fellowship program is needed to assist some African countries in this area. Particular attention should be given to arrangement of required training within Africa or in similar conditions.

External Assistance

Due to the financial crisis experienced by most African countries, there is a need for external funds, primarily for the creation of additional training capacities, expansion of existing facilities and the creation of centers of advanced studies.

REFERENCES

Food and Agriculture Organization (1984) Training Manpower for Agricultural and Rural Development. Rome: Author.

Food and Agriculture Organization (1984) Trained Agricultural Manpower Assessment in Africa. Rome: Author.

Food and Agriculture Organization (1984) FAO's Agricultural Training Activities in Africa. Rome: Author.

Food and Agriculture Organization (1984) Directory of Agricultural Education and Training Institutions in Africa, Rome: Author.

Chapter Sixteen

EMERGING PRIORITIES FOR DEVELOPING COUNTRIES
IN AGRICULTURAL EXTENSION*

Michael Baxter
The World Bank

Agricultural extension has received increasing attention from
both the governments of developing countries and development
organizations over the past decade. This attention in itself is
not new - many developing countries have devoted significant
budgetary and other resources to establishing commodity-or
area-specific extension systems in the 1950s and 1960s. What
is new is the scale and direction of this interest.

For many developing countries the importance of agriculture in economic development was reaffirmed in the 1970s.
This was also the time when considerable effort and investment were devoted to strengthening earlier-established agricultural extension systems. As well, the decade saw a
significant increase in the involvement of development organizations in agricultural extension. This interest of both
developing countries and development organizations (and of
academic institutions, in both the developing and industrialized worlds) in agricultural extension has continued to the
present. While today's interest should be seen against this
background, there are a number of emerging priorities for
developing countries in agricultural extension.

One cannot generalize on the status of extension worldwide or even on universal emerging priorities, given the
diversity of developing countries. The cultural, administrative
and agro-ecological conditions of each country are reflected in
local extension organization and extension priorities. Nonetheless, the experience of the World Bank with agricultural
extension in a large number of developing countries suggests
that there is perhaps one over-riding emerging priority -
efficiency. In particular, concern lies with how to increase
the efficiency of staff and other resources devoted to exten-

---

*The views and opinions expressed in this chapter do not
necessarily reflect the position or policy of the World Bank
and no official endorsement should be inferred.

sion as they address the fundamental objective of extension services - to increase the production and/or incomes of farmers they serve.

This paper examines four aspects of the efficiency issue: (1) the cost of extension, (2) links between research and extension, (3) technological development (particularly in communications technology), and (4) extension's work with women farmers. It concludes with general observations on how these emerging priorities relate to our fundamental concern of establishing effective agricultural extension services in farmers' fields.

COST OF EXTENSION

Concern with the cost of extension is not new. In establishing extension services, as much attention is normally given to cost as to any other factor. What is new is the realization that even the most stringent calculations of extension service requirements can require review, given the financial situation in which many countries find themselves at this time. There are four aspects of this concern that are of particular interest.

One is a realization that no matter how attractive and even successful an extension system may be on a pilot basis, it is of little relevance to the country at large unless it is replicable nationally in terms of organization and cost. The broader context must always be kept in mind. The number of field extension staff and of technical specialists are two factors which must especially be considered as programs are translated from pilot to national programs. Intensive coverage of farmers by agents supported by a relatively large number of specialists will, of course, give good results. The 'trick' is to establish a system that effectively balances the needs of the farming community with the number and quality of available (or readily recruited) staff and other resources.

A second aspect is the fact that extension is but one of a number of agricultural services that contribute to the success of agricultural development. Just as extension must be organized and managed effectively so, too, must these other services. In many countries, priority in strengthening agricultural services is often given to reorganizing agricultural extension. This cannot be done in isolation of other services, however, since effective extension depends as much on sound management of the agricultural ministry and the coordinated support of other services as it does in sound management of the agricultural extension service itself. Moreover, once it has been shown possible to reorganize the extension service, a variety of other functions of the Ministry appear redundant. What can begin as a relatively limited objective - to make the extension service more efficient - can soon lead to the realization that more fundamental organizational and procedural changes are required in agricultural (and, indeed, developmental) administration for reasons of both efficiency and economy.

A third aspect relates to the use of mass media. There are some genuine trade-offs that can be made between field extension services and mass media. They are not as great as many proponents (who appear to come primarily from other than developing countries) would have one believe, however. One reason for this is that farmers tend to prefer direct contact with extension agents in the field - especially to verify other sources of information, for assistance in pest and disease problems and for information on input availability. Another is that, in practice, there is considerable difficulty in coordinating farm programs on radio with the actual conditions of farmers, even where farming conditions are uniform over a considerable area (which they are often not). There is a role for well-coordinated media and field extension activities, but for most developing countries, the emphasis should be on coordination rather than substitution.

The fourth aspect is that there is justifiable concern in many countries over how to limit the cost of extension. Especially in environments of patronage and where state welfare systems are absent, extension can increase with little apparent relationship to the justifiable needs of the service. Not only should the size of the service be rationalized, but so should the proportion of funds allocated to personnel as opposed to operating (and to a lesser extent capital) costs. While many governments agree with such statements when they remain statements of principle, few have yet begun the zero-based bloodletting exercises that become increasingly required.

In order to establish systems of work that are widely applicable and to make the corresponding changes in administration, the effectiveness of extension investments must be evaluated from the perspective of the field. It is likely that some economy can be made in this case. Not infrequently, however, where economies are required (and then often in a crisis environment) the step taken almost by reflex is to limit operational funds. Rather than undertaking the deeper analysis needed for effective cost control in the long run, limiting operational funds just assures that whatever expenditures are made have little impact.

With the interest in the cost of extension, it is not surprising that 'privatization' and 'cost recovery' of extension services have taken on a certain attraction - and not only in developing countries. Given the fact that some privatization of extension services develops anyway, and with the generally acknowledged inefficiencies of bureaucracies, these concepts are even more attractive.

263

The greatest advances are being made in privatization. Private extension services are common even now in some developing countries. High-value crops frequently encourage the development of individual or small firms of private consultants who advise farmers on production and marketing of particular crops. Similarly, useful research and extension activities are often performed by input supply companies, though their proprietary interests are usually paramount. Cooperative societies built around vertically integrated industries (such as sugar milling in Maharshtra, India) often have their own 'extension staff' who perform a range of supply, advisory and marketing functions – much as the CFDT-inspired cotton and other commodity companies or BAT tobacco enterprises of Africa. These extension staff provide valuable services for the crops for which they are responsible. Rarely, however, are they adequately coordinated with government extension services operating in the same area. As well, they usually do not apply their resources to the food crops that are fundamental to the farmers who produce the cash crops upon which their attention is focused.

The World Bank is involved in a project in Chile where agricultural credit funds can be used to pay for extension advice from private individuals. Other projects, for example in Nigeria, have provided farm advisory staff to service larger-scale farmers with individualized farm planning and management advice. These advisory services have generally been on the public account, but the concept lends itself easily to privatization. Extension services provided by the private sector, or even profit-oriented parastatals, can only upgrade the quality of overall extension support available to farmers, both for the crops and activities they directly cover and by the competition they provide government. Similarly, one cannot argue against the principle of cost recovery. At the least it instills a sense of financial discipline.

There are limits, however, to the extent to which both privatization and cost recovery can usefully be pursued. Knowledge, the 'good' with which extension is concerned, is a public commodity. Unless knowledge is discrete and situation-specific, it is not well-suited to private transfer – hence the development of private advisory services for specific high-value, often high entry cost, crops rather than for, e.g. cereals and pulses. Also, a government is responsible for extension support to all farmers, many of whom in developing countries work in difficult environments with very limited capital and land resources. Considerable effort is required to ensure that the poorest and most isolated farmers receive adequate extension coverage, and that the extension service provides adequate feedback to research and other agricultural services. It is not clear how broad-based privatized extension services can fulfil these basic functions.

Although the issue of cost recovery is frequently addressed, there are few significant instances of its successful direct implementation, at least under conditions akin to those in most developing countries. One can argue that the historically urban-biased terms of trade typical of most developing countries more than offset the direct cost of government-financed extension services. When more quantifiable and, it would appear, a priori, enforceable systems of cost recovery are not enforced – most obviously water charges for irrigation, not to mention many urban-based welfare systems – the logic of insisting on cost recovery for extension is not readily apparent. Moreover, the difficulties in designing workable systems of cost recovery outweigh any foreseeable advantage.

RESEARCH AND EXTENSION LINKAGES

A second emerging priority of developing countries in extension is to improve the linkages between extension and research. The attention given this subject in developing countries (and elsewhere) attests to the realization of: (1) the need to strengthen the impact of extension on the identification and prioritization of agricultural research problems and to orient research to key significant problems faced by farmers; and (2) the fact that farmers, extension staff and agricultural researchers operate within the one system. The development of farming systems research and allied approachs is related to this general renewed awareness of the need for effective two-way communication between farmers, extension and research. The perspective on actual production conditions and constraints at the farm level that results from a farming systems approach can lead to unusual speculations on the role of research and extension vis-à-vis farmers [2], but the advantages of the perspective outweigh such problems.

A major difficulty facing many extension services is that the technology 'available' to extension workers is often not attractive to farmers. As a result of multi-disciplinary diagnostic surveys undertaken by farming systems research, a number of countries have begun comprehensive combined extension-research field reviews of farmers' production conditions and needs; examined the suitability of recommended technology; and determined extension and research 'gaps' in light of review findings. Interesting work is being done in this regard in Nigeria, and its useful action-oriented reviews have been done in India [3].

Such diagnostic surveys by extension and research to assess farmers' actual production conditions and needs may appear surprisingly fundamental considering the substantial funds already devoted to extension and research. Perhaps the most surprising element in this regard, however, is how often agricultural research and extension organizations are actually not in effective, professional communication. Even extension services that operate in the same zone as the agricultural research station responsible for that zone may have effectively no contact with that station. In such circumstances, extension attempts to propagate 'recommendations' long since heard and rejected by farmers, while research works towards optimal yields far beyond the interset and resources of farmers (even if they <u>were</u> to hear of the required technology). Under these conditions, any development - such as farming systems research - that brings farmers' conditions and needs into the practical focus of extension and research must be encouraged.

Related to the need for more effective identification by extension and research of farmers' key technological needs is the realization that experienced crop based extension services should also handle farmers' other production activities. Particularly once the methodological expertise of extension field staff is established and there is appropriate orientation of training and research programs, an extension service should pay attention to farmers' non-crop interests. The integration of livestock production (not veterinary services) with crop production within an extension service is an obvious step and one of basic importance in many farming societies, given the interrelations of crop and livestock production. In Indonesia where the initial orientation of the strengthened agricultural extension service was on crops (especially rice) the work programs of extension staff now includes tree crops, livestock and farm fisheries. In areas where there are active farm forestry programs, the need for extension's involvement in the field is obvious, and has been achieved in some Indian states. As staff resources permit (here quality is a more significant consideration than staff numbers) an extension service should become involved in all significant productive farm activities.

TECHNOLOGICAL DEVELOPMENTS IN COMMUNICATIONS

An emerging priority in developing countries is the effective utilization of technological developments in communications by extension. Three developments in communications systems and technology are particularly significant for extension: the proliferation of the electronic mass media (i.e. radio and television); the availability of small, handy video cameras; and the development of interactive video/computer systems. Each is reviewed below.

As noted above, if radio and television programs are closely attuned to farmers' needs and conditions, with particular attention to the timing of agricultural operations, they can be an effective adjunct to field extension services. Given the expansion of market-oriented agriculture and the increasing complexity of input requirements, radio and television can be powerful supports to field services, as well as to the on-going education of farmers and extension staff. However, there are considerable difficulties in achieving topical relevance, which is perhaps the main constraint to the effective use of radio and television in this sense in developing countries.

Small video cameras can have a significant impact on the quality of extension field staff training, on the quality of their support to farmers, and on the responsiveness of government to farmers. Small enough to be highly mobile and relatively unobtrusive, simple enough to be used with limited training, and capable of being played back on increasingly common VCRs, these cameras can dramatically narrow the gap between farmer and government, and between field, teacher and researcher. They have been successfully used in Latin America (for example in Mexico's Programa de Desarrollo Rural Integrado del Tropico Humido, PRODERITH) to elicit villagers' analyses of their development situations, needs and priorities. A more direct use of video cameras is by technical specialists, trainers and even extension agents to record crop conditions, etc. to stimulate both staff training and farmers. The advantages of flexibility in use and the field orientation they encourage outweigh the cost of these cameras in appropriate circumstances.

Advances in micro-computer technology have already had a significant impact on extension and on farmers' access to information. The development of interactive systems will undoubtedly further influence the quality of extension work. There are already advocates who would promote farmer access to such equipment. The most obvious use of such equipment for extension services in developing countries, however, is in the training of extension staff and to improve the relevance of technology development to farmers.

The quality of an extension services' technical staff, and consequently of its field staff, is often a major constraint to the upgrading of the service. Interactive video/computer systems could contribute significantly to overcoming this constraint. Interactive video-disc based training modules are one possibility. In India, for example, training modules on pump maintenance and irrigation water management are being developed. Uniformity of material and the self-pacing and constant evaluation of trainees are attractions of such systems - provided sufficient accurate and locally-relevant training modules are available.

A critical problem facing extension staff in many developing countries is that of having access to technical recommendations that are attractive to farmers, considering the farmers' particular agro-ecological and resource conditions.

Rather than being due to an absence of technology per se, this is often because research fails to take practical account of the resource conditions of each main type of farmer. Consequently, equally as important as the incorporation of interactive systems in training is their use in encouraging useful feedback from extension to research (and to other agricultural services) and the development of farmer-relevant technology. It should not be difficult to develop procedures to screen the practical suitability of technical recommendations both by researchers and by extension staff. Indeed, technically and economically, this would appear to be the priority in developing interactive systems to the advantage of extension.

EXTENSION FOR WOMEN FARMERS

How to better serve women who work as farmers is an emerging priority in many developing countries. The involvement of women in farming operations varies significantly between countries and cultures. It is not uncommon, however, for women to perform a greater share of agricultural tasks than men, and for technological developments (especially varietal improvements and improved implements) to benefit the crops or tasks more commonly dealt with by men.

The increased pressure on extension (and research) services to focus more effectively on the tasks performed by women is welcome. It has shown, however, that practical solutions are difficult to identify and even more so to implement. For example, the common proposal of having more female extension staff is not necessarily an answer if female employees are constrained from travelling freely and from meeting farmers, or if suitable technology for the tasks performed by women farmers is not available. At the same time, undue emphasis on 'home economics' by women extension staff only deviates attention from the critical issue of women's role in agricultural production.

Given the staff composition and orientation of most extension and research services and the poor track record of many 'women's extension' components, the task of developing effective extension for women farmers is not easy. Attention should continue to be given to the role of women in agriculture in particular societies and agricultural systems. Priority, however, should go to implementing small projects that utilize women-oriented extension methods that can be used both as learning devices and for awareness development in the village community, the extension service, governments and development organizations. Of the emerging priorities discussed here, the successful implementation of effective agricultural extension activities for women as farmers is likely to be the most difficult to achieve, but also perhaps the most significant in the long term.

The four emerging priorities for developing countries in agricultural extension discussed above all revolve around the theme of 'efficiency' - efficiency of resource allocation and of operation of the extension system. As fundamental as these priorities are, there is one basic principle that is perhaps so obvious it is frequently unstated, and so often overlooked. No matter the amount of investment, discussion and thought that goes into the design of an effective extension service, all that counts in the long run is how the system operates in the field. There comes a time when the after-all relatively uncomplicated central issue of extension - how to most simply and directly link the farmer and technology development in a two-way relationship - can no longer be profitably discussed.

The focus of governments, development organizations and academics alike must then be on ensuring that the staff, functions and components of the agricultural extension system each actively contributes to effective extension operations that directly benefit farmers. To return to the theme of this paper, then, the fundamental emerging priority for developing countries in agricultural extension is efficiency of operation: our work should contribute practically and simply to that end.

EPILOGUE

In summary, this volume is the product of several understandings: (1) that the extension function is carried out
worldwide through systems which are varied; (2) that these
systems are interdependent within the agricultural development process and are enhanced or benefit by linkages (at
several levels); and (3) that as systems they require
supports, especially policy and resource allocating supports.
A basic premiss of the volume is that these understandings
gain in depth through international review.

The first part of this epilogue is a discussion of selected
topics that have emerged primarily from the contributors'
perceptions presented in this book. These observations are
gathered under three main sections: I. The critical role of
policy in extension; II. Contemporary system practices and
issues associated with system operation; and III. The scope
and importance of linkages. The second section draws implications based on the above review.

I. THE ROLE OF POLICY

Agricultural extension, as noted many times herein, is an
interdependent system. While operating as part of the
research–extension–farmer continuum and as a specific
function in the larger agricultural development process, the
success of extension systems is highly dependent upon policy
– national policy, international donor policy, policy of other
related agencies, etc.

First, extension's success depends on national commitment both to the agriculture sector in general and to agricultural extension in particular. In the final analysis, the
question is whether public extension is considered to be a
priority in the complex array of national policies to ensure for
food security, exports and modernization in developing
countries.

A nation's commitment to agricultural extension is generally reflected in its development strategy. In LDCs in particular such strategies are often influenced by global politics and by international supports for agricultural development work - contributed by donor organizations. Thus, an external set of factors is usually involved in the official commitment and direction regarding extension - for example, in the cases where countries have adopted the World Bank-supported T&V system or where the USAID-supported FSR/E is in effect. Recently, there has been a trend toward international support for private enterprise, privatization and public-private coordination with respect to research and extension.

While professionals may argue that consistent policies must be promoted by government if there is to be any improvement made in the agricultural 'technical innovation process' and more specifically in agricultural extension, the reality is that public policy is shaped amidst controversies which affect priorities for development. For instance, while economic motives energize the interests of some, siding them with large producers and wealthier farmers - others uphold sociopolitical concerns i.e. questions of equity and the problems of the poorer members of society. As well, controversies change with changes in political administrations. These shifting debates affect long-range program effectiveness, in industrial as well as less-developed countries.

It is difficult to resolve policy issues, and some authors counsel (Johnston and Clark, 1982) that concentration be targeted on the techniques of development organization in order to avoid 'disembodied polemics'. But in reality these polemics are not separable from institutional and technical approaches to organizational development and program delivery. Confrontation and debate are inevitable - unless we disavow the role of leadership and declare ourselves to be only technicians.

One of the main challenges presented by the agricultural sector as a whole is that its study is truly interdisciplinary but paradoxically offers little reward for cross-disciplinary study and interaction. Decisions about extension, as with agricultural policy generally (Robinson, 1984) reflect the current technological, political and economic climate. However, a narrow perspective tends to prevail. In the policy arena, for instance (especially in times of budget retrenchment), what would normally be systems and program development questions are typically taken up by politicians, with technicians and planners left with diminished influence.

This volume underscores various policy issues relating to extension which continually demand attention: issues regarding economic and social goals, public vs private roles, issues of centralized vs decentralized planning strategies, and issues

271

relating to incentives, target audiences, and general extension system design. A brief review of these issues follows.

Overall economic policy issues, as Schuh emphasizes
herein, are critical in shaping the agricultural process in
developing countries. He cautions that the economic policies of
LDCs have a tendency to discriminate against the agriculture
sector and favor development of the industrial sector. Indeed,
policies regarding pricing, credits, inputs and marketing can
make or break production programs, but extension services
inevitably receive the blame - even when research is inadequate or inappropriate! A major issue, then, involves the
overall commitment of LDCs to socio-economic development in
agriculture and whether they are willing to coordinate overall
economic policy to benefit extension systems.

Although public policy is primary for agricultural extension, an important contemporary policy issue highlighted in
this volume relates to public policy goals for private sector
agricultural development and coordination between the public
and private sectors. Rodgers addresses the contemporary
issue of whether private agricultural extension bodies may be
alternatives to the public system. While advocating the
importance and value of fostering private enterprise development in the agricultural extension domain, policy makers
must note that this contribution can only supplement the
public effort - not provide a substitute for it. Indeed, the
public effort must pursue agricultural extension purposes
somewhat distinct from the clear profit motive of private
firms. The public/private is thus not an either/or proposition
- but a question of cooperation and coordination.

A public policy issue with both political and technical
implications relates to participation through decentralization.
Rondinelli presents a typology of four kinds of contemporary
decentralization - deconcentration, delegation, devolution, and
transfer to non-governmental institutions. This move
represents a major policy shift in the international arena,
which heretofore had pursued centralized strategies of development. The turnabout from the nationally centralized to
decentralized efforts can be traced to the mid 1970s and a
backlash in the West with respect to the efforts of international organizations to encourage developing countries to
plan more systematically at national levels and to encourage
central planning in general.

Moris, as Onyango, encourages national policy makers to
note the relationship of appropriate incentives for farmers and
agents to successful agricultural extension. With respect to
farmer incentives, the discussion again turns to policy
questions of price ratios between sales and inputs, and to
credit and input subsidies. Regarding agent incentives,
especially in sub-Saharan Africa, Moris reviews problems
posed by inadequate technical packages, the lack of co-

ordinated bureaucracy, the feedback of messages 'from below', and untenable working conditions.

A target-audience type of sociopolitical issue has remained a 'sleeper' until recently - the role of women as agriculturalists. Weidemann takes an economically-oriented stance on this issue by stressing the advancement of agricultural production through recognizing the role and importance of women in farming. She advocates knowledge generation that takes stock of the problems faced by women farmers, as well as knowledge transfer that is targeted toward women farmers.

Policy issues often come to the fore when countries begin to make choices or consider reforms regarding the 'ideal' extension system. What is to be the character of the system? What approach should the system follow - commodity focused, community development-cum-extension, technical innovation centered? Who is to benefit? Who will control and regulate the system? How will the system be operated? By whom? - i.e. what agency, set of agencies or coordinated bodies, public and private? Most would agree in principle that system design should not be 'one-way', but balance 'top-down' technology transfer with 'bottom up' farmer-indicated need. The issue is how to arrive at that balance and what the costs will be. Axinn, for example, makes the distinction between delivery systems (usually commandeered by Ministries of Agriculture) and acquisition systems (organized by farmers). Recognizing that there are also hybrids of these two types, he points up that the type of system employed reveals both who controls it and the intended beneficiaries.

The very fact of differing systems and practices of agricultural extension leads to issue-taking and results in particular policies and practices sponsored by donor agencies and/or countries. For example, United States assistance efforts attempted in the 1960s and 70s to adapt the Land Grant System to situations in developing countries, only resulted in disillusionment. Subsequently the official stance of the U.S. Agency for International Development turned toward development of the Farming Systems Research and Development (and Extension) approach and, more recently, toward enhancement of communications technology and advancement of private 'contract farming' schemes.

Public policy issues also arise consequent to major developments. At the international level considerable resources have gone into the development of a global system of international agricultural research centers (IARCs). The challenge of determining the return on this investment and the value of investing in international vs national research has raised considerable discussion. Evenson's study helps to demonstrate the impact of international research on the advancement of national research and extension, and reflects the internationality and interdependency of the world food and agriculture production systems. It must be noted, however, that by the mid-1980s the Consultative Group for International Agricultural Research, CGIAR, is 'experiencing increasing difficulty in obtaining support for its budget - at the very time that the system is reaching maturity and beginning to produce a steady flow of technology' (Schuh, 1985, p. 5). In part, this development appears to be due to the impatience of policy makers who hope to witness major changes in development in the short-run when, inevitably, these changes require long-range and continuing effort.

To discuss public policy is to deal with shifting issues, changing political administrations, continually changing values, and distinct priorities that emerge over time. The value of agriculture itself, as well as the value afforded to production extension and knowledge transfer are constantly in flux. But underlying the inevitable debates and the discontinuities in policy development, this volume suggests that the extension function must continue to be performed and that there exist certain basic features common to effective extension systems.

II. SYSTEM PRACTICES AND ASSOCIATED ISSUES

Once public policy has dictated the importance of agricultural extension, the next step is to design effective systems to meet the articulated objectives. It is hoped that as policy makers begin to design or modify extension systems for their respective countries they will benefit from the comparative perspective of this book.

Within this volume, varying opinions on agricultural extension system effectiveness prevail. Benor advocates the highly managed T&V approach to extension, exhorting officials to reconsider the originally formulated key aspects of the T&V system (which are often 'misunderstood or ignored'), and to 'return to the basics' in developing T&V and T&V-derived systems.

McDermott argues for the newly emerging Farming Systems Research and Extension approach that views the research-extension-farmer continuum as parts of a total technical innovation process (TIP), recognizing that the extension agent is the prime interface between the TIP and the farmer. He proposes that there is (and should be) no functional separation between research and extension. Elsewhere it is suggested by Denning (1985), and herein by Moris and Rivera, that other 'hybrid' systems combining selected features of various systems may be needed, e.g. in certain areas such as sub-Saharan Africa.

Discussions herein have also centered on questions of systems (and, indeed, policies) as they apply to specific countries. Onyango's instructive review of the case of Kenya provides an example of how one country has dealt with the policy issues of decentralization, incentives, and public/private cooperation as it has moved toward reinvigorating its extension system. Decentralization has been put into practice in Kenya through the District Focus (Kenya, 1984), whereby national planning and budgetary boards make their decisions on the basis of expressed District needs. New incentive systems have been established in both the public and private sectors. These incentives include promotion, professional career development opportunities, and allowances for public extension agents; and input guarantees and price reviews for farmers. Also, the private sector has been encouraged to cooperate more closely through linkages between voluntary organizations, cooperative societies, banks, private business, and industry. Farmer participation has been particularly stressed through agricultural extension efforts (FTCs, RTCs, demonstrations and farm tours) and through the farmer organized body known as ASK - the Agricultural Society of Kenya.

But national level policy makers should also be aware that the application of models that are successful in one country may not necessarily guarantee success in another. Blum, in recounting Israel's experience in agricultural extension, is particularly cogent. He observes that, while there are certain principles that must guide the development of extension systems, there are particular features underlying any successful system that cannot necessarily be transferred. These include: a vision of what research and extension can accomplish in a particular country, an attitude of dedication to the system, and a readiness to be inventive in the variety of situations that continually arise in the office and in the field which require spontaneous ideas and fresh approaches. Thus, regardless of system, there are leadership, cultural and other factors which remain eminently important for success.

The book continually highlights the importance of good management to successful extension systems. Rivera recounts the recently accelerated move toward agricultural extension management training programs for top officials, a development being spearheaded by India. In reviewing agricultural extension developments in India and in particular those related to the T&V system, he notes (as does Roberts) that agricultural extension is dependent on various aspects of the agricultural development process. He proposes that extension management training include personnel from various agencies: viz. credit institutions, supply agencies, research bodies, production extension services, marketing agencies, as well as in the technical areas. He argues further that extension management training should involve an emphasis on the development of administrative competencies related to specific extension systems.

While good management is essential it does not mean, however, that effective leadership is necessarily in place. As organizational experts know, one manages things but leads people. Even more importantly, as Blum notes, a vision must prevail so that enthusiasm and inventiveness can flourish. A system is only a framework within which leadership and management can take place.

Thus far we have noted two primary arenas - macro and micro - where various factors help determine the success (or failure) of agricultural extension: (1) policy preferences, national circumstances and agricultural development planning strategy at the macro level; and (2) the effective design and organization of the extension system itself, including the efficiency of its officers at the micro level. Between these two, but related to both, is the concept of linkage. Generally in the extension literature, 'linkage' refers to the research-extension-farmer continuum (the technical innovation process). In the following section, which is based on the presentations in this volume, this definition is expanded to include other critical linkages.

III. THE SCOPE AND IMPORTANCE OF LINKAGES

In this volume, the importance of linkages between extension and other arenas of agricultural development is a predominant theme. Indeed, Roberts concludes that the context may frequently be more important than the type of system employed, and that this context must consist of at least the following four key factors: an agricultural research network with linkages to extension, credit and input supply systems, farmer incentive structures, and effective use of government extension staff.

Linkage along the research-extension-farmer continuum is critical. Without effective interaction along this continuum, extension could be thought of as a truck speeding without lights (no policy direction), without cargo (no technology), and without a clear destination (no targeted clientele). The burden of this interaction rests as much with policy makers and researchers as with extension professionals.

Although research-extension-farmer linkages are crucial to the 'technical innovation process', to quote McDermott, other linkages remain significant for success. They include political linkages that move up and down in principle from policy to agency to farmers and other clientele and back, system linkages that extend across the entire agricultural development process, and scientific-technical linkages (which move along both the larger system of the agricultural development process and the sub-system of the research-extension-farmer continuum). In this last illustration, we begin to see the various entities within the agricultural development process more accurately as a system of agencies operating interdependently, not separately. In sum, political, system and scientific-technical linkages all appear to be required in making agricultural extension fully effective (Figure E.1).

Another vital linkage, as Maalouf points out, is critical if the agencies and organizations involved in the agricultural development process are to be supplied with qualified personnel - the linkage between the agricultural education systems of formal education and pre-service training with the

I.   The Research-Extension-Farmer Linkage

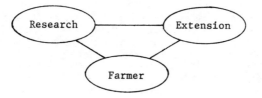

II.   The Agricultural Development Process Linkage

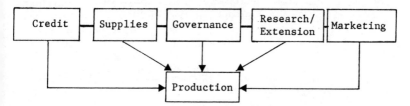

III.   The Political Linkage

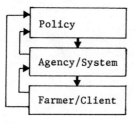

Figure E.1:   Three types of linkage for agricultural extension

nonformal agricultural extension system. Underlying the entire agricultural development process is the need to train qualified professionals at the intermediate and higher education levels. Maalouf reviews the findings of two FAO studies, one on training institutions and the other a manpower assessment in Africa, Saharan and sub-Saharan. Notable is the fact that some countries have many institutions while others at best have one or two. Thus there appears to be a need for cross-country interchange, for the effective regional utilization of existing institutions, for strengthening existing training programs, and for linkages between training and research institutions.

Thus, while agricultural extension operates within a triangulation generally known as the research-extension-farmer continuum, its interconnections are wider (Axinn and Thorat, 1972). Agricultural extension needs to maintain effective, on-going linkages with policy makers as well as other departments, agencies and related services. The time is long overdue when it should have emerged as a priority. What appears to be required is the provision of linkage management skills aimed at fostering interagency action as well as the installation of institutional arrangements to ensure for co-operative planning so that extension can meet the stated objectives formulated for it by national policy makers. Such linkages would contribute to the achievement of national planning goals, cross-sector cooperation, and the advancement of new paradigms in extension's development.

IV. CONCLUSIONS

In conclusion, it appears that contemporary decisions regarding national agricultural extension must be made with particular reference to international and system interdependency, and possibilities for regional cooperation. As well, when designing or redesigning extension systems governments must take into consideration: national vision and will, available qualified leadership, required organizational improvements, effective linkages, and the importance of farmer feedback.

In addition, this volume suggests that aside from the practical problems of research/extension development, certain conceptual problems must be confronted. These conceptual concerns involve the advancement of creative paradigms and insightful experimentation.

IV.A. PRACTICAL PROBLEMS

Interdependency

Interdependencies operate globally as well as along the continuum of the agricultural development process. The global perspective of this volume emphasizes the many international challenges regarding extension.

Ironically, at a time when United States farmers and agricultural leaders are retreating from international commitment, we are entering a period when more international concern is required. As Schuh (1985) points up, developments in the international economy have changed the context of agriculture within national economies. Indeed, agriculture today in LDCs as well as in modernized countries depends in large part on exports, and therefore on the international market. This is a new reality for countries such as the United States which has significantly expanded its export of agricultural goods since the 1950s. International markets and opportunities are affecting what extension should know and what it does. Farmer incentive programs, policies regarding supplies, the environment and markets - all influence production and profitability, and thereby impact on the relationship between extension professionals and farm producers.

The international agencies have wielded a strong influence among LDCs and have significantly impacted upon agricultural extension's development. The United Nations System - including the World Bank, the FAO, UNESCO - and various bilateral agencies, especially the United States AID programs, have been concerned with extension and have affected the choices of LDCs as to its development. In addition to policy determinations, these organizations have contributed much to institution building and program initiation through technical assistance. International organizations have also provided arenas for discussion and program action, stimulated innumerable research projects and publications, and generally contributed one after another landmark to the development of the fields of agricultural extension and adult education. We take the position that a long-term and consistent commitment to these organizations is needed, that they should be more widely recognized for their contributions to international development.

Regional Cooperation

The growing interdependence among the nations of different hemispheres and continents is one of the most important trends of recent years. In some cases this interdependence is still underdeveloped, as in sub-Saharan Africa. Maalouf points up the findings of the FAO studies which show that geographically adjacent nations often differ radically in their institutional resources. One idea for alleviating these imbalances is greater regional cooperation.

Shared projects for technology generation and transfer appear to be an emerging priority which signals the opening of a promising avenue to regional trade, as well as implementation of joint projects through binational and multinational agreements (in Africa as elsewhere). The problems and potential facing countries in similar regions and comparable circumstances provides an opportunity for these countries to set standards collectively to govern action programs undertaken, thereby assuring sustainable development. Too often individual countries find themselves deferring to the needs assessments of donors. This may result in only partially accepted pilot projects which terminate when donor support is withdrawn. As well, regional commitment and cooperation among countries may further development effects in their interests in technology transfer from industrialized countries.

National Vision and Will

A primary principle in generating and organizing agricultural knowledge and its transfer is that of national commitment, or national will. Improved agricultural production and farmer profitability begins with a stated and supported policy and a national plan for the economic, social and political goals of a country. Generating (or testing for the applicability of) technology and its transfer also requires specific goals - goals that can be met and evaluated within realistic time frameworks. Often, goals are too ambitious or time-frames unreasonable.

National will is an imprecise concept, but becomes even less exact in federally constituted countries where major national decisions take place on a nationwide basis through the collective determination of states, or provinces. India is an example of a country where the states are undertaking a single (albeit slightly modified from state to state) approach to agricultural extension - the T&V system. This state (or nationwide) commitment is supported by the union government so that we can speak in general of a 'national commitment' to this particular system.

Whatever the complexities and imprecision of the concept, national will is evidenced, as already noted, in stated policies and supported plans. These policies must cover provisos for the broad incentives for development as well as the specifics of agency and program, including management. Again India is an example - where the Prime Minister has called on all government executives and personnel to upgrade themselves through periodic training in management skills.

Qualified, Motivated Leaders

There is also need for qualified and motivated leadership within agricultural extension systems around the globe if technology is to be effectively and efficiently transferred. At present, this would appear to be a low, rather than a high, priority - in the United States as elsewhere. From the FAO studies in Africa, assessing the agricultural education and training situations and their needs, it becomes apparent that it is important to educate and train agriculturalists for improved systems development. Curriculum with a broader perspective needs to be developed for these individuals - providing training to analyze social, economic, political and educational forces, and to enhance clear understanding of extension organization. Also agents and program leaders require training in systematic program development, but with a broader focus than is usually addressed. The success of extension systems depends on understanding the external forces which affect extension's responsibilities and structural development, and also on internal advancement through promotion of management skills and training opportunities for individual improvement.

Organizational Improvements

Organizational improvements and innovative institutional arrangements are needed at the public agency level, particularly in LDCs. Problems of bureaucracy, poorly qualified and untrained personnel, inadequate structural arrangements, insufficient linkages, etc., plague the public sector. Incentives in extension work, an area highlighted in this text, are often lacking. Institutional development also tends to fluctuate as a priority. Certainly, development tends to have its 'fads,' but human resource (human capital) development must remain a major and ongoing commitment.

Linkages and Linkage Management

Although stressed earlier in an earlier section of this epilogue, it bears repeating that one aspect of management seldom considered is that regarding linkages. Linkage management, as already mentioned, involves at least three major arenas: (1) the research-extension-farmer continuum, (2) the political linkages between the extension institution and enabling (resource-granting) organizations, and (3) the system-wide linkages with agencies and systems across the agricultural development process as a whole. This particular view of linkages has yet to come into 'vogue,' and thus lacks priority status, but at the basic continuum level it does not take genius to recognize that those extension systems which operate with success are those where research and extension closely interact.

Farmer Feedback

The critical point in any discussion of agricultural extension is whether it is effective at the farm level. It is only when known knowledge (effective methods) and new knowledge (research) have moved from the basic to the applied to the adaptive level that we may say that knowledge has any meaning. This top-down movement requires the assistance of extension - no matter how or by what institution that extension service is provided. However, for that knowledge (the known and the new) to be truly applicable it must be fitted to the needs of the farmer, and the best way of insuring for this is to insure for bottom-up farmer feedback. In the most successful extension systems we note that farmers impact on the extension and knowledge-generation systems both administratively and through feedback regarding research directions.

IV.B. CONCEPTUAL PROBLEMS

New Paradigms

There is a need for new paradigms and practices in the development of extension systems and programs, as well as in thinking about the role of knowledge transfer and its importance across the entire agricultural development process. Indeed, the insights emerging from this volume suggest several integrative conceptual frameworks regarding agricultural education, extension and the agricultural development process.

First, consider the term 'technology transfer' to be narrow as well as inappropriate in its suggestion as to the needs of certain LDCs. We argue that a more neutral and inclusive concept for the agriculture sector as it refers to its 'educational' and 'transfer' activities is that of knowledge transfer - whether that transfer occurs in formal, nonformal or informal settings and whether it involves new technology or simply techniques known to be effective.

Second, the term 'extension' has also become bothersome. One tends to think of 'extension' as a service carried out only by agricultural extension systems. In practice extension is a function undertaken by most agencies and systems in agricultural development. As well, while referring to systems which often have quite distinct purposes (commodity production at one extreme and community development at the other) and numerous approaches - in the context of less-developed countries, the term extension has come to be narrowed and used only to mean 'production extension' activities (Baxter, 1986). This emphasis tends to limit discussion of extension internationally to its function for production purposes within the agricultural development process.

As a result of ingrained language barriers, as well as a certain inertia, the field has tended to be stymied. In the United States for example, a modern nation where the industrialization of agriculture is a fait accompli it is only with the contemporary criticism of the extension mission - that renewed concern about that mission has developed.

As already noted, the agricultural development process itself ranges over such a large conglomerate of businesses, industries, and production-related activities that the overall 'system' of public agencies which operate to assist in making the process work is often considered only in fragmentary analysis, with a focus on one agency - usually the ministry or department of agriculture. The complexity of the system requires a broader view and more inclusive concepts. If the goal is an effective agricultural development process, we must break through ingrained institutional and conceptual boundaries and view the process as a whole.

The systems approach is a useful organizing tool. It is a way of thinking about a problem and structuring an analysis in terms of the major issues, the critical variables and the linkages among variables and among sub-systems. The inter-dependencies of these sub-systems affect the overall system performance and consequently are most relevant from the perspective of higher policy levels. It is such an approach that we envision in this discussion of the knowledge transfer concept.

Our argument for the integrative concept of 'knowledge transfer' is based on its conceptual range. It appears to overcome the segmenting problems of sectoral distinctions (e.g. education vs agriculture), the limiting tendency to organize the concept of extension around production only, and the impractical lack of interconnectedness among the various agencies involved in the agricultural development process.

The concept of knowledge transfer may also help to highlight the agricultural development process as a system - since it includes input, conversion and output agencies operating independently but interdependently functioning within the agricultural development process. Knowledge transfer is a term which need not be limited to any particular agency or organization. The agricultural development process - with its many business connections and its equally profuse number of public agencies involved - needs knowledge to be transferred and shared within the entire system, and continually out to the primary clientele: farmers. Indeed, knowledge as to price policies, credit, supplies, and markets is continually transferred to farmers - not only knowledge for production purposes. By conceiving of the agricultural development process as a whole requiring knowledge transfer throughout the process, from every agency - enabling, functional, technical, and knowledge generation and transfer agencies, new approaches to agricultural extension development may emerge.

As this volume makes evident, the factors affecting extension's success refer to policy and agency, as well as diffusion and adoption. For example, let us recall the main factors that have contributed to Kenya's success: (1) the structuring of agricultural extension to provide administrative and technical support; (2) attention to the organizational structuring of public services which control development strategies; (3) incentive systems for extension workers and the clients of extension; and (4) a philosophy of extension within the national development effort. From a different perspective, we note with Roberts that there are at least four key extension requirements for success: availability of improved technology, supply of inputs, credit availability, market infrastructure, and favorable government pricing policies. These are frameworks within which any 'factors for success' must operate.

Experimentation

In addition to new paradigms and broader perspectives, we would agree with those contributors who underline the need for, and importance of, experimentation in agricultural extension development. Experimentation is needed with respect to research methodologies, field practice, and the development of extension systems. Areas of experimentation mentioned herein include integration of existing systems (such as T&V and FSR/D), development of hybrid management systems based on contingency analysis and preferences of local decision makers, and models of public/private cooperation.

New practices in agricultural extension, including those mentioned in this volume, are indicative of new paradigms. These new paradigms appear to be coming into play as professionals begin to conceive of new ways of thinking about extension. Indeed, it has been one of the purposes of this volume to bring conceptual insights to practical problems and by the study of practical problems to develop insightful concepts. It is our hope that we have succeeded in this and have thereby contributed to efforts to improve extension and advance agricultural development.

REFERENCES

Axinn, G.H. and Thorat, S. (1972) Modernizing World Agriculture: a Comparative Study of Agricultural Extension education systems. New Delhi: Oxford and IBH Publishing Company.

Baxter, M. (1986) Speech in Minnesota on T&V, Paper. Washington, DC: The World Bank.

Denning, G. (1985) 'Integrating Agricultural Extension Program with Farming Systems Research'. In: Cernea, M.M., Coulter, J.K., and Russell, J.F.A. (eds), Research-Extension-Farmer: A Two-Way Continuum for Agricultural Development. Washington, DC: The World Bank.

Hage, J. and Finsterbusch, K. (1985) Organizational Change and Development: Strategies for Institution-Building. College Park, MD: University of Maryland, Dept. of Arts and Sociology.

Johnston, B.F. and Clark, W.C. (1982) Redesigning Rural Development: A Strategic Perspective. Baltimore and London: The Johns Hopkins University Press.

Kenya. Office of the President (1984, Rev. June) District Focus for Rural Development. Nairobi: Govt. Printer.

Ladaga, F. (1986) 'Development of an Extension System for Philippine Cocoa Smallholders'. Dissertation. College Park, MD: University of Maryland, Department of Agricultural and Extension Education.

Lowdermilk, M.K. (1985) 'A System Process for Improving the Quality of Agricultural Extension', Journal of Extension Systems (New Delhi), 1 (Dec.), pp. 45-53.

Lynton, R.P. and Pareek, U. (1978) Training for Development. West Hartford, CT: Kumarian.

McClelland, D.C. (1965) 'Toward a Theory of Motive Acquisition', American Psychologist, vol. XX, pp. 319-33.

Ogden, D.M., Jr. (1984) 'How National Policy is Made'. In: Taylor, L. (ed.), Change. NY: Cooper-Hewitt Museum (Smithsonian Institution).

Robinson, B.H. (1984) 'Fifty Years of Food Policy: What Have We Learned?' Paper presented at the annual meeting of the American Association for the Advancement of Science, New York

Schuh, G.E. (1985) 'Strategic Issues in International Agriculture'. World Bank paper (draft) dated 4/4/85. Washington, DC: The World Bank.

INDEX